D1544534

The
MEAT BOOK

The MEAT BOOK

✦◗◆◗

A consumer's guide to selecting, buying, cutting, storing, freezing, & carving the various cuts

✦◗◆◗

by

TRAVERS MONCURE EVANS

and

DAVID GREENE

with illustrations by

DANA GREENE

CHARLES SCRIBNER'S SONS / NEW YORK

Text copyright © 1973 Travers M. Evans
Illustrations copyright © 1973 Dana J. Greene

This book published simultaneously in the
United States of America and in Canada—
Copyright under the Berne Convention.

All rights reserved. No part of this book
may be reproduced in any form without the
permission of Charles Scribner's Sons.

1 3 5 7 9 11 13 15 17 19 V/C 20 18 16 14 12 10 8 6 4 2

Printed in the United States of America
Libraray of Congress Catalog Card Number 73-1337
SBN 684-13565-5 (cloth)

for David and Richard

Preface

❁❁❁❁❁

It was on an April evening in 1968 that I first met David Greene. The place was Harlem. The occasion was a meat consumer seminar. As the audience took their seats in what was formerly the ballroom of the old Theresa Hotel on 125th Street, David Greene, who was to lead the seminar, and his assistant were busily setting up the props that would be used to illustrate the evening's topic.

The props, of course, were meat, meat of all kinds: whole ribs of beef, whole racks of lamb, fresh and smoked pork tenderloin, bacon, hot dogs, chopped meat. There were, in fact, examples of almost all the familiar cuts and some of the not-so-familiar.

My attitude about the upcoming performance by David Greene was jaded at best—perhaps even a little condescending. Okay, Mister, I'm pretty well versed when it comes to the kitchen. Anything new on your butcher block? Show me.

And show us he did! His knife twinkled as it deftly cut and trimmed away. David Greene had an uncanny way of doing the totally unpredictable to an ordinary piece of meat and magically transforming it into three sumptuous meals for a family of four. The audience was enthralled and gasped in admiration as the handsome cuts of meat were passed around for inspection. As he lectured and cut away, I began to get his message. David Greene was showing us how to *understand* meat in almost the same way that a highly knowledgeable butcher does. And more importantly, he was teaching us how to apply this understanding to our own day-to-day meat consumption habits, and most importantly to our pocketbooks.

Since that eventful evening, I have attended many of David Greene's seminars. All have been informative, imaginative, and lively.

David Greene's career began as a salesman for a major meat-cutting processor. During World War II he taught the first live meat-cutting course ever attempted in the army, and following his tour he once again returned to the business of buying and selling meat.

About fifteen years ago his own frozen meat concern made its appearance: The Greenway Meat Service. Greenway offered a highly successful line of gourmet frozen meats with an emphasis on beef. The carefully wrapped packets, complete with cooking instructions, could be found in the freezer cases of many exclusive retail markets and in the food section of department stores under the "Hotel Brand" label.

Although Greenway's meat was Mr. Greene's primary source of income, for years his primary source of concern had been the consumer. The various radio shows, television appearances, and the consumer seminars he instituted and conducted in communities where he felt consumer protection was greatly needed helped in a small way to get his message to the people. But Mr. Greene dreamed of sharing his knowledge and experience with more people. He wanted to clear up the confusion about meat in order to make consumer power a truly potent force. *The Meat Book* is the culmination of that dream. It encompasses the vastness of Mr. Greene's knowledge, explores his authoritative ideas, and expounds on meaty facts that he believed every consumer should know. His knowledge and his enthusiasm pervade these pages. Not one is untouched by his philosophy. His expertise on meat enabled him to stand unchallenged by this statement: "Any cut of meat will be tender if you buy good quality and cook it properly."

It would have been David Greene's proudest moment to see *The Meat Book* in print. But that was not to be. In January 1972 he suddenly passed away.

Now, with fine, generous help from a great many people, *The Meat Book* is a reality. For this, special thanks are due four very special men without whose help this book might have forever remained a manuscript in limbo. Thanks also to David Greene's daughter, Dana, who illustrated these pages, Pete Fischer who first saw virtue in the idea, Johanna Lee Murray for endless hours of typing, David Adams for sharing his own meat-cutting expertise and knowledge with the author and illustrator, and lastly my husband, Charles Evans, for unrelenting encouragement.

T.M.E.

Contents

❈❈❈❈

PART VII Consumer Power

Cook it right!

	Thickness	Total cooking time	Instant-reading thermometer temperature
Rare	1 inch	15-20 minutes	140°
	1½ inches	25-30 minutes	
Medium	1 inch	20-25 minutes	160°
	1½ inches	30-35 minutes	
Well	1 inch	25-30 minutes	170°
	1½ inches	35-40 minutes	

NOTE: Cooking times and temperature readings are for warm-weather barbecuing. For an accurate temperature reading, place instant thermometer as pictured, *left*.

Carve it right!

● Always use a very sharp, 8- to 10-inch knife; it will help you carve like a pro. ● For thin, tender slices (like the ones pictured, *left*, from a Top Round Steak), cut at a slant and across the grain of the meat. Carving meat on a slant also gives you slightly larger serving sizes. ● Be unconventional and creative. Instead of just placing chunks of a steak on each person's plate, divide steaks by their sections (such as a Porterhouse Steak) and carve each section in slices. You'll usually get more servings this way, too.

Sirloin Steak

Sirloin Steak ($1.99-$3.29/lb.): This top-rated steak is excellent grilled. There are four different-looking Sirloin Steaks, which can be identified by the shape of the bone: pin or hip bone, flat bone, round bone and wedge bone. A meaty first choice would be the "flat bone" steak (*pictured*), because it contains a full-size

● Trim off excess—but not all—fat to avoid "flare-ups," which can cause unnecessary burning of meat.
● Cut slits in the fat that rims the steak to prevent it from curling up.
● Brush steaks that are not marinated with a light coat of oil to prevent them from sticking to the grill.
● If the steak isn't marinated, pat it lightly with a paper towel to remove

THE OUTDOOR CHEF'S GUIDE TO STEAKS

There's nothing quite like the aroma of a steak sizzling on the grill. But how do you pick a great steak? At today's high prices, you can't afford a "lemon." Follow my guide and you'll make no mistakes. By EMILIE TAYLOR

When you get to the meat counter, do you often find yourself confused, wondering which steak is really the "best looking" and the best buy? To help you,

First Cut Chuck Blade Steak ($1.29-$1.79/lb.): This can be a wonderful—and tender—barbecue steak if you remove the coarse-textured strip of meat, called the cap, from the top of the steak. What's left is a hearty but tender steak. There are four types of first cuts. You can tell the most tender one because it has a straight white bone near the top section of the steak. Marinate, if desired. This 3 lb. steak serves two to three.

Center Cut Chuck Blade Steak ($1.39-$1.89/lb.): Before you barbecue this meaty steak, remove the very soft grainy section of meat on the top right side of the steak. Grind for great "steakburgers." There are four different-looking center cuts. Look for the one with the bone that peaks up slightly on the right or left (depending on how it's packaged). Marinate, if desired. This 3½ lb. steak serves four.

Chuck Top Blade Steaks, Boneless ($1.99-$2.99/lb.): These tender Chuck Steaks are often overlooked. Pick the steaks with the finer line of white tissue going through the center. When grilled, these little steaks are surprisingly tender. Marinate, if desired. A 3 oz. steak serves one.

Chuck Shoulder Steak, Bone-

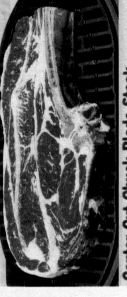

First Cut Chuck Blade Steak

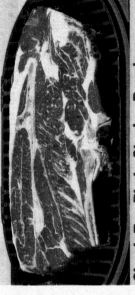

Center Cut Chuck Blade Steak

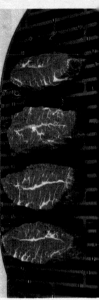

Chuck Top Blade Steaks, Boneless

Chuck Shoulder Steak, Boneless

Rib Eye Steaks

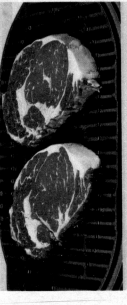

Porterhouse Steak

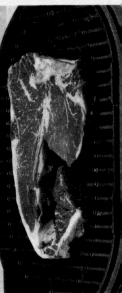

BILL McGINN

ture guide of the 10 best beef steaks.

• **First, check the price.** Always buy steaks on sale, which usually means saving from 50¢ to $1 per pound. Next, look at the color, fat and marbling.

• **Color:** Top-quality beef steaks should be light red.

• **Fat:** Beef fat should be firm and white, not gray or yellow.

• **Marbling:** Those little flecks of fat running through the meat are what's called marbling. Marbling keeps meat moist while it cooks and gives steaks a tasty, juicy flavor.

• **The drier the package, the better.** Excess liquid loss means a less tender steak.

• **Study the 10 types of steaks pictured on this and the following page.** We've chosen the most desirable cuts you should look for—for instance, the best First Cut Chuck Steak, the best Top Round Steak. If these preferred cuts are not available, other cuts of the same type steak are also good, if the quality and price are right.

HOW TO PREPARE A STEAK FOR GRILLING

• Trim off any white tissue around meat. ⟫→

this steak for barbecuing, but with a little marinating, it's a great family feeder. This lean, moderately tender steak has a heavy diagonal line of tissue going through it. Since this cut sometimes contains two or three sections, look for the most solid piece. Remove tissue after cooking and before carving. Must marinate. This 2½ lb. steak serves four.

Rib Eye Steaks ($3.99-$4.99/lb.): These exceptionally tender boneless steaks are delicious when grilled. It's the fat in the meat that gives this cut a very rich flavor. Also called Club Steaks. This 1 lb. steak serves one to two.

Porterhouse Steak ($2.79-$3.49/lb.): A barbecue-lover's delight, this very tender cut contains a "T"-shaped bone that actually divides two higher-priced steaks. The top section is called the Top Loin, Shell or New York Strip Steak; the section under the bone is known as a Filet Mignon or Tenderloin. Many Porterhouse Steaks, like the one pictured, also contain a "tail," which can be left on the steak and barbecued, or ground up for hamburgers. This 1½ lb. steak serves two to three.

EMILIE TAYLOR learned to cut meat from her father, an expert butcher. She runs the Emilie Taylor Meat School in Paramus, New Jersey.

Sirloin Tip Round Steak

Top Round Steak

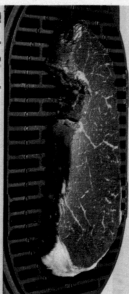

Flank Steak

This 2¾ lb. steak serves four to five.

Sirloin Tip Round Steak ($1.99-$2.99/lb.): Another tasty choice for barbecuing, this boneless steak is moderately tender. It contains a fine line of tissue going through one side of the steak. Don't confuse this cut with the more tender Boneless Sirloin Steak, but don't overlook it either. It's economical and satisfying. Must marinate. This 2½ lb. steak serves five.

Top Round Steak ($1.99-$2.99/lb.): Also called a London Broil, this is a favorite boneless steak for barbecuing. There are three different-looking Top Round Steaks. Look for the most tender one—it's uneven in shape. Butchers call it the First Cut or Oyster Cut. Marinate, if desired. This 2½ lb. steak serves five.

Flank Steak ($2.99-$3.99/lb.): This popular barbecue steak has come up through the ranks, not only in position but in price. Be sure to slice this very lean, thin, grainy piece of meat across the grain. It gets thicker when cooked on the grill. Marinate, if desired. This 1½ lb. steak serves four.

• Let meat stand at room temperature for one hour before cooking. Heat will penetrate faster so the steak will need less cooking time.

• Don't let frozen meat "overthaw." This causes a loss of natural juices, which takes away from the tenderness and flavor of the steak.

BARBECUE IT RIGHT

• Get those coals good and hot, until they're glowing. Preheat a gas grill for 10 minutes before cooking.

• Place steaks that are 1-inch thick from 2 to 3 inches from heat; thicker steaks should be placed from 3 to 5 inches from heat.

• Try not to turn a steak more than once; frequent turning causes a loss of natural juices and can make a steak tough.

• Use tongs instead of a fork to turn meat. Puncturing meat causes loss of natural juices.

• Never cut into the lean of the meat to check for doneness. If you must, cut along the bone or between the sections of the meat. An instant-reading thermometer is the best test for a rare, medium or well-done steak. If you don't have an instant-type thermometer, use our "cooking timetable," *above*.

■

The MEAT BOOK

Part I

Introduction

An illustration from an Egyptian tomb showing the preparation of meat.

Chapter 1

Meat Mystiques and Economics

curious superstitions, taboos, rituals
customs, fancies ... and facts about buying meat

The earliest evidence suggests that man is basically a carniverous animal. Charred bones found in caves tell us that meat was a diet staple of our oldest ancestors. Down through the ages meat has played a curious role in primitive ritualistic ceremonies, figured in superstitious literature, been forbidden under taboos that still exist today, and served as the ultimate status symbol on our tables. Probably no food has been more respected, more suspected, so esteemed or so abhorred. Small wonder considering some of the strange beliefs surrounding it.

Before the birth of Christ, Roman priests burned the entrails of cattle in sacrifice to their gods, but saved the rest of the beef for themselves to gluttonously feast upon. And in the eighteenth century, in some circles it was widely believed that eating meat was a major cause of cannibalism.

In an African tribe, young girls refuse to eat pork for fear that their children will look like pigs. But broth made of fresh pork was used as a remedy for many maladies in nineteenth-century Scotland.

Horace wrote that liver was widely used as an aphrodisiac during his time. And in Shakespeare's *Twelvth Night*, Sir Andrew confesses: "I am a great eater of beef and I believe that does harm to my wit." Indeed, until very recently, the widely held belief *was* that large quantities of meat would make one belligerent.

One of the strangest pagan rituals was performed in worship of Demeter, the Greek goddess of the grain. To help the crops along, the people sacrificed and ate pigs, but saved some of the pork to sow with the seed at the next festival. The pork they feasted upon in this particular rite was believed to be a holy sacrament, symbolic of the flesh of the goddess herself.

Although most of us in America are unaware of it, a curious physical phenomenon actually does occur from eating meat fat. Eating meat fat produces butyric acid, and butyric acid causes a person to have a distinctive body odor—a distinctively unpleasant one according to those who never eat meat. Until the end of the last century, the Japanese, who were largely fish eaters, could hardly abide the scent imparted by their meat-eating Western neighbors.

Some other widely held superstitions stemming from ancient times are these: meat is a stimulant; it will make one virile; it is bad for one's kidneys; it shouldn't be eaten in hot weather because it will cause one's body temperature to rise. All are mythical.

Meat was so strictly forbidden during the Lenten seasons of the early Christian church that violation of the laws resulted in stiff penalties. A statute passed in England in 1570 recommended a fine of 60 shillings and imprisonment for three months for all violators. There were certain exceptions however: the sick, aged, and infirm. But even those people had to have the legal authorization of a physician in order to include meat in their diets.

Even at that early date, it seems, doctors had an inkling of how important meat is to the diet. Meat, along with eggs, is recognized today as our most valuable and easily digested source of protein—and protein is essential to growth and life. In fact, many nutritionists recommend several portions of meat be eaten every day—which may seem unrealistic, what with today's prices—especially if one has a family to feed and a tight budget. Which brings us to the purpose of this book. We believe that it is possible, even today, to purchase meat economically—if it is bought with expertise. Although we don't believe that there is really any such thing as an "economy cut," and we know there are many of us who cannot afford the price of filet mignon, there is a way to get the meat you need in your diet and your money's worth at the meat counter.

For starters, a consumer with meat in mind should be acquainted with the factors that determine meat's value—characteristics such as these: (1) grade; (2) the amount of trim required; (3) the cooking method called for; (4) the number of servings per pound and the nutritive value per serving.

GRADE

Prime, Choice, and Good (or in the case of pork, #1, #2, and #3) identify the U.S. Department of Agriculture grade of the meat. The higher the quality of the grade, the tenderer, tastier, and juicier a piece of meat will be. And high quality means higher cost.

Meat graded Prime is the highest in price and quality. Choice grade meat is second highest and the most popular and widely available grade of meat on the market today. All the cuts of meat described and discussed in this book are USDA Choice grade. Meat graded Good ranks third in quality and is not as widely available as Choice grade meat.

TRIM

The amount of trim required of a piece of meat will depend to a large extent from where it is cut. When you purchase steaks, roast, or chops the options are usually threefold:

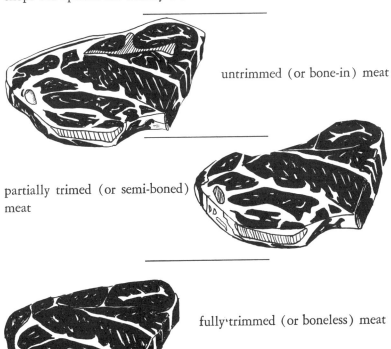

untrimmed (or bone-in) meat

partially trimed (or semi-boned) meat

fully‧trimmed (or boneless) meat

Most people believe that the more cutting, trimming, and shaping a cut of meat requires, the higher its price will be. After all, meat cutting reuires a salaried expertise—as much as 6 cents per minute. Strangely enough, however, fully trimmed cuts are *not* more expensive necessarily, because what you are purchasing is practically 100 per cent meat. Purchasing that untrimmed, bone-in cut at the lowest end of the price spectrum can be false economy. The hidden costs are the extra bone, fat, and gristle in every economy priced pound. Most of it will end up in the garbage pail. The higher priced, fully trimmed piece of meat will have far less waste and enough solid meat to justify the price differential.

Beef chuck is a perfect example of the "hidden cost" theory. A bone-in, 2-pound chuck steak at $1.19 per pound will probably have about the same amount of solid, lean meat as a 1½-pound, semi-boned chuck steak at $1.37 per pound, and a 1½-pound, fully trimmed chuck steak at $1.49 per pound. Your cost will actually be about the same for all three, and each will deliver about the same amount of meat.

The same theory can apply to the lower cost veal, pork, and lamb chops labeled "end cut." They have more bone, more gristle and fat, and less meat than the higher priced and infinitely more desirable center-cut chops. Center-cut chops are meatier, tastier, and more tender.

COOKING METHOD

The way a cut of meat will be cooked also accounts for a cost differential. Generally, the cuts that can be roasted or broiled will cost more than cuts that must be pan fried or moist cooked. The reason is that roasting and broiling are "dry" methods of cooking. Meats that can be roasted or broiled have a high degree of inner marbling. Marbling provides the natural moisture required to keep these cuts juicy and tender as they roast or broil. Marbling also guarantees the meat will cook more quickly because it helps to conduct the heat. Well-marbled meats also offer a high degree of cooking flexibility. Anything you can roast, you can also pot roast, and anything you can broil can also be pan fried. But the reverse is not true. Because they contain so little marbling, most of the less expensive cuts will require long, slow moist cooking—braising, pot roasting, or stewing—to make them tender enough to eat. But the less tender cuts can make great meals with time and a little cooking ingenuity. And you can often save pretty pennies by cutting those lower-cost cuts into cubes yourself for a variety of moist-cooked meals.

SERVINGS PER POUND AND NUTRITIONAL VALUE

How much meat will be enough to feed each person? Again, this depends on the cut. (See Chapter 18, "Selecting Self-Service Meat," for a chart.) You will probably need to buy more of the "low-cost," high-waste cuts of meat than you will of a filet mignon that gives you total meat and no waste. If a cut contains no bone and very little fat, as little as three ounces of meat can suffice per serving. If you are serving something like beef plate meat, breast of lamb, or spareribs, however, you are getting very little meat and lots of bone. You may need upwards of a pound per serving.

To judge how much meat to buy when you are selecting a cut, gauge the amount of fat to meat, bone to meat. Depending on appetites, somewhere between one quarter and three quarters of a pound of meat per serving is the usual amount you will need.

Something to avoid is the lower-priced package labeled "chopped meat" or "ground meat for cookouts." It could contain an astonishingly high percentage of fat. This fat will partially cook out as the meat is heated, and the hamburger patties will also shrink away. In spite of its higher cost, you are better off buying ground chuck or ground round at 15 cents to 30 cents more per pound. It has less fat, so will shrink less. And it will provide more protein nourishment.

Bacon offers yet another example of extreme shrinkage during cooking. How many times have you put a skillet full of bacon on the range to cook, only to return minutes later and find it shriveled down to nothing and floating in drippings? What you are paying for in bacon is mostly fat, and that is money down the drain. Sliced smoked butt cut from the pork shoulder and fried is a tasty alternative accompaniment to breakfast eggs. Although it is priced about 10 cents more per pound than bacon, it offers more than enough lean meat to compensate for the difference in price. Serving for serving, smoked butt is a better value than bacon.

So you think hot dogs and luncheon meats are thrifty? Consider this fact if you do: most hot dogs are made up of 10 per cent water, 30 per cent fat, a small percentage of filler (unless the label reads "all meat") —more than 40 per cent non-nutritive foodstuffs. You would have to eat three hot dogs—or up to fifteen slices of luncheon meat—to provide the protein nourishment in just three ounces of lean beef.

Most families spend 33 per cent to 50 per cent of their food budget on meat and meat products, yet spend very little time learning the dos and don'ts, whys and wherefores of recognizing quality, buying, cutting, and carving the various cuts of meat economically. The following pages give you these straight facts in simple, graphic, easily understood descriptions and illustrations in an attempt to clear up the confusion and misconceptions surrounding the purchase of meat.

Revealed here are secrets your butcher never told you about sirloin, short loin, brisket, and chuck—all manner of beef, pork, veal, and lamb. You'll learn how long meats last and how to make them stretch further, just how much to allot per portion, how to carve cuts properly, maybe even why that roast you cooked last night was stringy, dry, and tasteless. We will also show you how two 2- to 3-pound center-cut sirloin steaks, as an example, can be the makings for three or more complete meals for your entire family. Those makings would include more than a pound of beef kabobs, four small filets, and eight to twelve sandwich steaks. Soup is even included in the bargain if the bones are put to good use.

In fact, before you know it, you'll have the satisfaction of fileting steaks, tying your own roast, the knack of recognizing quality and value and exactly what you're buying, and how to cook it for maximum tenderness.

There may be no real economy cuts of meat, but it is still possible to save money on the meat you purchase. Here are some hints that may be helpful:

1. Watch your weekly specials. They can save you up to 20 per cent or more on your overall meat expenditure. (See Chapter 18, "Selecting Self-Service Meat.")

2. Learn all the meat-cooking methods. Properly cooked meat will always be tender. This book matches the cooking methods with the cut of meat. (But we leave the creative cookery up to you.)

3. Serve home-cooked meats rather than "convenience" meals.

4. Get a sharp knife and learn how to bone, trim, and make special cuts of meat at home. (See Chapter 20, "The Step-by-Step Guide to Cutting Meat.")

5. Know your meat terminolgy. Consult Appendix 1 at the back of this book if in doubt.

With the knowledge of the variety of cuts available (and a little imagination) there is no reason why the vast supermarket selection—or your butcher—need ever be intimidating again. So read on and learn how you can buy better, cut better, and—better yet—*eat* better meat.

❖❖❖❖❖

A BRIEF MEAT BUYER'S BETTER-VALUE CHECKLIST

The next time you plan to serve that thrifty pork sparerib dinner to the crowd, consider this: pound for pound, serving for serving, a standing rib roast will cost you just about the same amount of money. (And it is a far more elegant and impressive dish.) Just to prove the point that there are no economy cuts of meat, the following further examples are cited:

LEAST COST	BEST BUY	REASON WHY
Beef		
bastard roast (sirloin)	bottom butt sirloin	easier carving, less waste
beef plate for stew	beef chuck for stew	more meat, less bone
front-cut brisket	straight-cut brisket	even slices, more lean meat
London broil (chuck)	London broil (top round)	less fatty tissue, delicate flavor
pinbone steak	Porterhouse steak	less bone, more tender meat
plain rib roast	oven-ready rib roast	easier to carve, less waste
rib steak	boneless club steak	less waste
round steak	loin steak	more marbling, quicker cooking
rump roast	top round roast	less fat, more tender meat
Pork		
end-cut chops	center-cut chops	less waste, more meat, flavor
hocks	whole shoulder	less skin and bone, more meat
rib-end chops	loin-end chops	less waste, more meat
rib-end pork loin	loin-end pork roast	less waste, more meat
spareribs	pork chops	less waste, more meat

LEAST COST	BEST BUY	REASON WHY
Ham		
bacon (sliced)	smoked butt	no waste, all meat
picnic ham	full butt	more tender, flavorful meat
semi-boned ham	boneless ham	less waste
Veal		
breaded veal cutlets	veal scallops	all meat, no bread
breast of veal roast	boned shoulder of veal	no waste, more meat
shoulder chops	loin chops	less bone, more tender
veal chops	veal cutlets	all meat, no waste
Lamb		
breast for stewing	neck for stewing	more tender, flavorful
shank for stewing	shoulder for stewing	more tender
shoulder chops	loin chops	more meat
shoulder roast	leg of lamb	less waste, better flavor

Part II

Beef

Hereford

Chapter 2

Quality Beef (and how to recognize it)

production... grading... selecting quality beef

In the U.S. we eat more beef than any other meat. The average person consumed more than 115 pounds of beef in 1972, and in spite of meat boycotts and rising prices the demand for beef continues to grow. It represents a gigantic capital outlay. How can the thoughtful, thrifty beef buyer get the most for his money? The answer to that question is—buy quality. What determines quality beef? Let's look at the meat pictured on the jacket of this book. The fat is discolored, the marbling scanty, the bones look crumbly. In the picture it has poor conformation and finish. All these aspects are important in considering the overall quality of beef.

The U.S. Department of Agriculture has set up a grading system to classify meat according to quality. Under this system, the USDA quality or grade of beef is stamped on each major cut. Whether the names of these grades—Prime, Choice, and Good—actually serve the consumer or confuse him has been the topic of a recent discussion by a study group in the House of Representatives. USDA Good, for example, is only third best, while the most popular grade, USDA Choice, is merely second best, and USDA Prime is best. Nevertheless, until the names are changed to Good, Better, and Best, the present system, which is optional but set up by the U.S. Department of Agriculture in cooperation with meat producers, is the only one we consumers have on which to rely. In spite of its present shortcoming with regard to specific nomenclature, this system can and should be relied upon.

To further protect the consumer, the U.S. Department of Agriculture makes it mandatory for meat in interstate commerce—which represents most of the meat that we eat—to be federally inspected for wholesomeness and cleanliness before it is shipped to markets.

U.S. Department of Agriculture stamp (certifying wholesomeness and cleanliness)

The round USDA stamp indicates that the meat is thus approved. The number on the stamp identifies the establishment which processed the meat. If there were any irregularities in the meat, presumably a consumer could track down the producer through this number. The other USDA stamp certifies the quality—or grade—of beef, which will be further discussed in this chapter.

In some instances, ungraded meat is sold by retail stores and freezer meat dealers. The reasons for this could be threefold: the dealer does not believe in the USDA system; he does not want to pay the grader's salary; or possibly he wants to make up his own meat grades. Ground meat cannot be graded because it is usually a blend of various grades as well as imported meats.

Under the USDA system, a federal agent stamps both the round certification of wholesomeness and the grade of beef on the whole carcass with a purplish colored vegetable dye. As the carcass is cut, these marks are usually trimmed away—and further labeling is often necessary at the point-of-sale.

The grade of beef is determined by many factors. Modern technology and agricultural methods are constantly upgrading and standardizing the quality of beef. The beef we eat today is simply better than ever before.

On the ranch, quality is produced by highly controlled scientific methods. Young cattle are put on fenced-in pastures and left to graze until they reach one year of age. They are then brought to confining feed lots and given grain food and various food concentrates. Because of their relative immobility, the cattle fatten, but their muscles do not develop further, and after about six months they yield a tender, high-quality grade of beef. This supervised method of cattle raising is specifically designed to upgrade the quality of the beef and make it rate the coveted USDA Prime or Choice stamp. If you are looking for top quality, these are the grades you should buy.

There are no rules or regulations governing the grade of meat a merchant may sell. Very often the skillful meat cutter will combine the ground meat trimmings from USDA Prime or USDA Choice grade meat with those of USDA Good. This practice abounds in the preparation of hamburger, meat loaf, and sausage stuffing, and indeed improves the overall quality of the product.

There are five government-approved grades of beef:

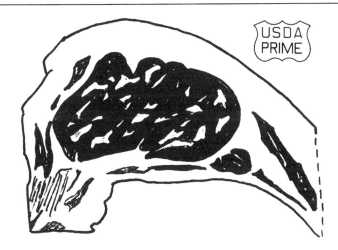

1. **USDA Prime**—This is the highest quality and the most expensive beef. It comprises only 10 per cent of the beef cattle grown in this country. USDA Prime beef is the beef from well-fed, young cattle. It is not widely available in self-service supermarkets except by special order because most of it is purchased by restaurants, hotels, and the more exclusive retail outlets whose trade demands and can afford it. The high internal fat content and thick, wasteful outer layer of surface fat make it about 30 per cent more costly than the next best grade. The fat does contribute to the superb flavor of the meat, but not to its nutritive value.

What to look for: A professional meat buyer looks at the general confirmation and finish of the animal. Confirmation refers to the animal's structure and shape. Finish has to do with the fat. Prime grade meat is "well finished." It has a high-quality, well-developed, evenly distributed fat content—both externally and laced through the internal lean meat. This high fat ratio promises the buyer that the meat will be tender, juicy, flavorful, and suitable to all cooking methods.

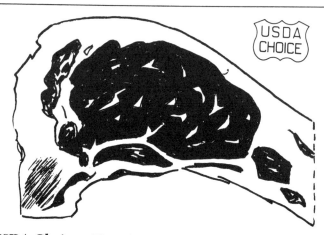

2. **USDA Choice**—There is more USDA Choice grade produced in this country than any other grade of beef. It is widely available in both supermarkets and butcher shops. Choice grade is very desirable, and contains less wasteful exterior fat than USDA Prime grade beef, but enough interior marbling to guarantee a tender, tasty, juicy piece of meat. This is the most popular beef with consumers because it is thriftier than Prime and suitable to all cooking methods. All meats discussed in this book are of this category.

What to look for: Choice grade beef is one of two types, heavy Choice or light Choice.

Heavy Choice is close to USDA Prime grade. It contains more fat and interior marbling than the light Choice, and weighs a little more. It is a very tender quality of beef that yields the better steaks and roasts that can be cooked by any method. For most consumers, however, it will take a nearly professional eye to distinguish between heavy Choice and light Choice.

Light Choice is a good buy for thrifty shoppers because it costs slightly less per pound than heavy Choice but has a comparable flavor. It is not as tender because it contains less fat. It is best, therefore, when pan broiled, fried, or moist cooked.

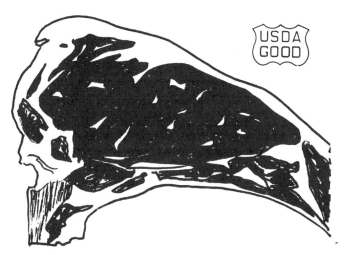

3. **USDA Good** is not as widely available as USDA Choice, although it is still the major consumer selection at some supermarket chain stores. This grade of beef lacks the fine flavor and interior marbling of Choice grade and consequently has a greater tendency to shrink and dry out during the cooking process. USDA Good is often ground in combination with USDA Choice beef trimmings for hamburger meat.

What to look for: Very little fat. This meat is best adapted to moist-cooking methods, or with the addition of fat, to pan broiling.

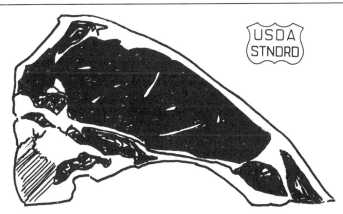

4. **USDA Standard** is rarely available in markets and not very suitable for home consumption. This grade of beef has less flavor, and if available should only be considered by shoppers who prefer very lean

beef. It is cheaper than the other grades listed above and requires long, slow, moist cooking.

What to look for: A sparse amount of fat both outside and within the lean meat.

5. **USDA Commercial**—Most of this quality beef is purchased by meat processing companies for sausages, frankfurters, and cold cuts. The illustration shows a thick outer layer of fat which is a result of the fact that much of the cattle from which Commercial grade beef is cut is older, more "rangy," and not suited for home consumption. Clever restauranteurs, however, have learned how to capitalize on steaks cut from the loin of USDA Commercial beef by cooking them with the aid of tenderizing agents and intense heat. Because this grade of beef is so inexpensive, they can offer their clientele a "complete steak dinner for $1.98." The only suitable cooking method for this grade of meat at home—regardless of cut—is moist cooking.

Other important aspects of beef to consider before you buy are:

Color—High quality beef is a bright, rosy red. Its fat is creamy white. Slight variations in this coloration may imply a more mature, sinewy cut of beef. Generally, the brighter and deeper the color, the younger the beef. Prepackaged supermarket meat is sometimes gray due to moisture condensation and a long shelf life. This discoloration means that some of the juices and freshness have been lost and the beef has lost its "bloom." It is still edible, though lower in nutritive quality, and if purchased at a price reduction should be eaten the same day.

Shoulder arm cuts
arm bone

Leg or round cuts
leg or round bone

Shoulder blade cuts

blade bone near neck

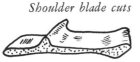
blade bone center cuts

blade bone near rib

Rib cuts
back bone and rib bone

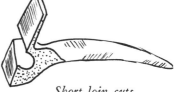
Short loin cuts
back bone (T-bone)

Sirloin cuts (hip)

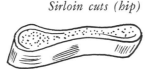
pin bone
(near short loin)

flat bone (center cuts)

wedge bone (near round)

Brisket cuts
breast bone

Bone structure—If the bone in a cut of beef appears knobby or round, it indicates the cut comes from a less tender part of the animal such as the shoulder or round. The more tender cuts such as sirloin, rib, and Porterhouse have flat bones.

Marbling—The easiest way to spot a high-quality piece of beef is by its interior marbling in combination with bright, rosy-colored lean meat. Marbling is the term used in reference to the delicate webbing of interior fat, resembling a pattern in marbled stone. The marbling interlaces the lean meat and adds juiciness, tenderness, and flavor. Some people ill-advisedly presume that the leaner the piece of beef, the more tender it will be. Exactly the opposite is true. Marbling helps retain natural moisture within the meat while it cooks. The amount of marbling actually spells out the ultimate flavor, juiciness, and tenderness of the beef, and also gives a hint about how to cook it.

You can usually follow this rule of thumb: the greater the degree of marbling, the higher the quality of beef. The only exception to this rule is overly mature beef. Sometimes it is highly marbled. However, one can easily differentiate a mature piece of beef from a younger one by its lack of rosy color. The following is a description of the marbling in the various grades of beef:

USDA Prime is very heavily marbled throughout. The marbling on a cut surface of the meat is evenly distributed—resembling little flakes of snow.

USDA Choice has adequate marbling, but it is less evenly distributed than that of Prime grade meat. Choice grade beef's marbling has a delicate "lacy" look.

USDA Good has spotty, moderate marbling that looks somewhat like grains of rice.

USDA Standard and Commercial grades of beef are not recommended for consumer steaks and roasts. They lack sufficient marbling to make them as suitable for home consumption as the other grades.

A question that should be considered at this point is that of aging beef. Aged beef is sometimes available in exclusive retail outlets and it is very expensive. The question we raise is this: is aging necessary and does it, in fact, contribute vitally to the overall quality of the product?

The aging of beef is generally accomplished by a meat wholesaler who chooses primal cuts of USDA Prime or at least USDA heavy Choice grade beef. (Lean cuts do not age satisfactorily because they lack the thick outer layer of fat that protects against spoilage.) The cuts selected for this old fashioned method of aging are generally ribs, short loin, and shells. They are tagged and stored at slightly above the freezing point for a period of two to eight weeks. During the aging period, as the subtle changes in flavor occur, bacterial growth appears and natural moisture evaporation causes the exposed surfaces of the meat to become dry and discolored. Does this beat fresh meat? It is strictly a matter of taste. Naturally, the fuzzy ends of the meat where the bacteria has sprouted will ultimately be trimmed away. But the consumer pays for it, and pays dearly for the trim, the shrinkage, the waste, and the process itself. Since the meats selected for aging are of the highest quality, it seems to us that the ever-so-subtle changes produced by aging are simply not worth the money.

Certainly aging is a process that should never be considered at home. The temperature of household refrigerators is not low enough to retard bacterial growth to the extent required. Neither should aged cuts of beef be frozen. They will simply lack the sweet taste of fresh meat, and their fine texture will break down and become pulpy rather than tender. Supermarkets and small local stores very rarely stock aged beef. The meat they sell is bought with the aim of a fast turnover.

Most meat today is somewhat aged. Before it reaches your plate it is already one to two weeks old. The exception is Kosher meat which by religious law must be eaten within seventy-two hours after the animal is slaughtered. But the newest packaging techniques by which meat is stored and shipped in vacuum-sealed polyethelene bags make it possible to age meat without shrinkage or visible deterioration for three weeks or even up to a month.

Chapter 3

The Anatomy of a Beef

positioning the primal cuts for tenderness

J uiciness, tenderness, and flavor—those are the characteristics for which a piece of beef is esteemed! The best way to predetermine how juicy and tender a piece of meat will be is to know the area of the cattle's anatomy from which it is cut. This is also the best clue as to whether the meat will be lean or marbled, tender or sinewy, juicy or dry, and whether you will have to roast or moist-cook it, broil or fry it.

There are seven basic cuts from a side of beef. They are known as the primal cuts. All of the primal cuts are composed of beef muscle, bone, fat, and connective tissue. The extent of muscular development will be determined by the location of the primal cut. Generally, the least well-developed muscles which yield the cuts of beef that can be roasted or broiled are located in the internal area of the primal cut. The leanest and most sinewy beef is cut from the more highly developed muscles which are usually located nearer the exterior of the primal cut. These cuts are most likely to require moist cooking.

Every rule has its exceptions, however, and in the case of beef, the exceptions are the rib and the loin cuts. These cuts come from muscles located high on the animal's back. Unlike the lean and muscularly fibrous beef shoulder (or chuck of beef) which is comprised of muscles that are mobile, the rib and the loin "ride" the animal and are basically made up of immobile muscle which is less developed and contains very little connective tissue. Therefore, the rib and the loin of beef are the tenderest sections and in great demand by the consumer.

A simple way to see this in your own mind's eye is to compare a side of beef to your own anatomy. Your shoulder, for instance, is constantly mobile—made of muscles that help you to carry, to push, to pull, to throw. The muscles in your ribs and your back, except for support, are carried along by the momentum and mobility of the muscles in your legs. Because these muscles "ride," they have, as a rule, fewer of the sinewy fibers and less muscular development than the more mobile parts of your body.

By law most states require that the anatomical location—the primal cut in other words—of the piece of beef you buy be clearly indicated on prewrapped packages. London broil is a good example. The generic term "London broil" was originally applied to a very special cut. The first and true London broil was cut from the inside flank of the hindquarter of beef. Today when you find it in markets it is labeled "London broil (flank)"—and delicious when broiled or moist cooked and carved very thinly on the bias. (Contrary to popular opinion, this method of carving does not tenderize meat, it merely doubles the width of the slices—and helps to preserve some of the juices of the meat.)

Because butchers and meat departments everywhere are marketing London broil cut from the round, shoulder, chuck, and top sirloin, they are required in most states to label the package with the identifying anatomical primal cut. Consumer advocates, however, have long deplored this practice and claim it is confusing and misleading. It need not be, however, if you recognize that London broil (round) or London broil (chuck) is not London broil (flank)—but ordinary round or chuck steak cut to a double thickness. Either cut will actually top the tenderness of flank when properly cooked.

As you can see, knowing the origin of a cut of beef is basic to your overall understanding of the subject of meat.

Side of Beef

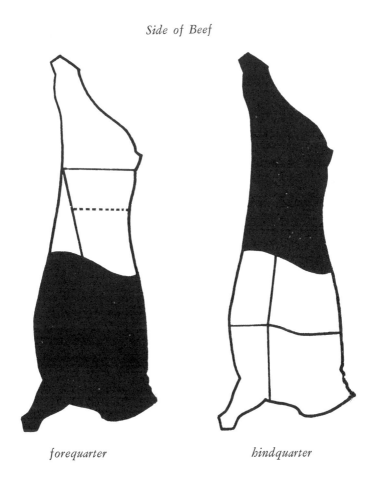

forequarter   *hindquarter*

DIVIDING A BEEF

An entire beef carcass weighs between 600 and 750 pounds before it is divided into two sides to be marketed. Each side of beef is then divided into a forequarter and a hindquarter. From these two portions come what is called the seven primal cuts. Four of the primal cuts come from the forequarter and three from the hindquarter. Learn these seven primal cuts and you will know many of the terms in a meat cutter's jargon at your butcher shop and supermarket meat counter.

THE SEVEN PRIMAL CUTS

The two forequarters comprise 52 per cent of the carcass. It is from these quarters that all Kosher meat is cut. A forequarter weighs from 155 to 190 pounds and is divided into four primal cuts:

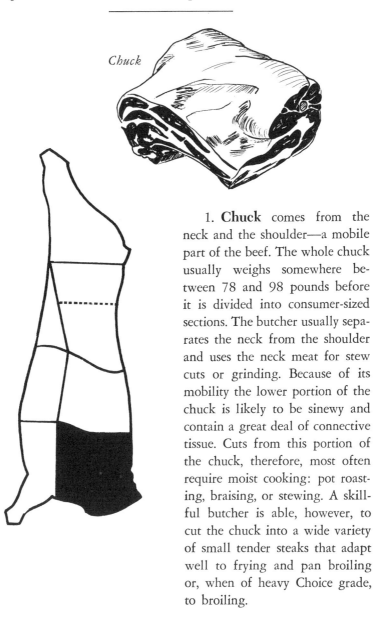

Chuck

1. **Chuck** comes from the neck and the shoulder—a mobile part of the beef. The whole chuck usually weighs somewhere between 78 and 98 pounds before it is divided into consumer-sized sections. The butcher usually separates the neck from the shoulder and uses the neck meat for stew cuts or grinding. Because of its mobility the lower portion of the chuck is likely to be sinewy and contain a great deal of connective tissue. Cuts from this portion of the chuck, therefore, most often require moist cooking: pot roasting, braising, or stewing. A skillful butcher is able, however, to cut the chuck into a wide variety of small tender steaks that adapt well to frying and pan broiling or, when of heavy Choice grade, to broiling.

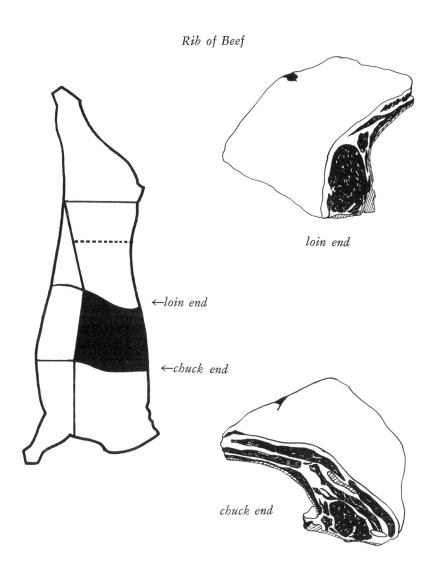

Rib of Beef

loin end

←*loin end*

←*chuck end*

chuck end

2. **Rib of beef** is the only immobile primal cut of the forequarter. It yields the most tender steaks and roasts—those that are most eminently qualified for roasting or broiling. A Choice grade primal cut of beef rib should weigh between 28 and 40 pounds.

Brisket

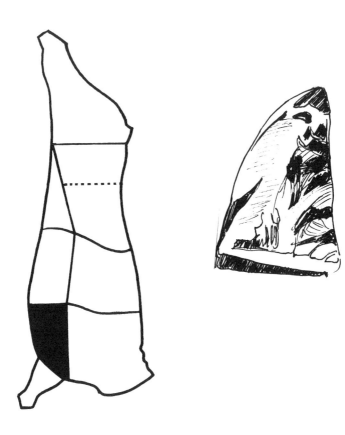

3. **Brisket** is a very fibrous part of the beef with lots of connective tissue and fat. This tasty cut comes from a highly mobile exterior muscle of the animal and will require either long, slow moist cooking and/or curing to make it tender enough to eat. Cured brisket is known as corned beef.

Short plate

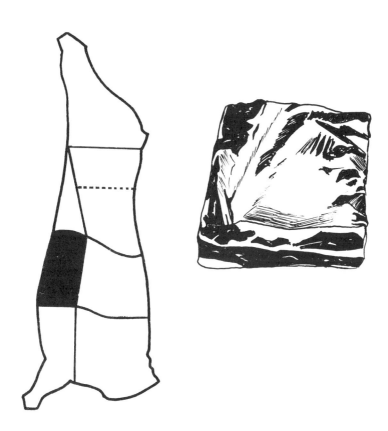

4. **Short plate** is part of the breast bone, the short end of the brisket. With the exception of the skirt steak, which is often referred to as an inside tenderloin, the cuts obtained from the short plate are adaptable only to moist cooking or grinding. This meat is usually combined with the lean parts of the chuck for grinding hamburger meat.

The hindquarters comprise 48 per cent of the steer carcass, but yield more steaks, more roasts, and less stew and chopped meat than the forequarters. A hindquarter weighs from 145 to 180 pounds and is divided into three primal cuts:

5. **Full loin** is the most select section of the hindquarter, the tenderest part of the entire beef carcass, in fact. It is the primal cut from which the beef tenderloin roast and the Porterhouse, sirloin, and filet mignon are cut —the roast and steaks that taste best and are most popular. The full loin is in great demand. It is subdivided into the *short loin* and the *sirloin* (also called the hip).

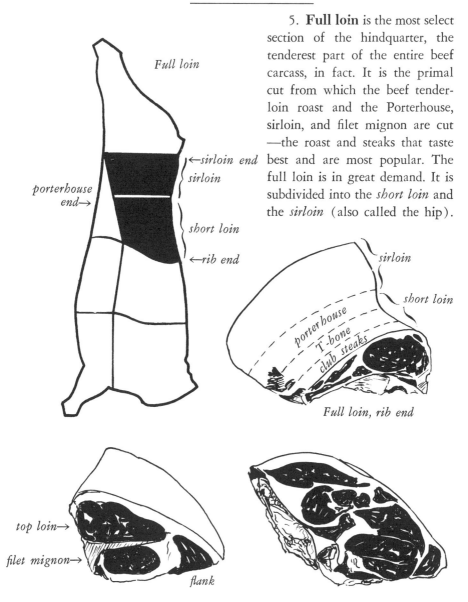

Full loin

←sirloin end

sirloin

porterhouse end→

short loin

←rib end

sirloin

short loin

porterhouse

T-bone

club steaks

Full loin, rib end

top loin→

filet mignon→

flank

Short loin porterhouse end

Full loin sirloin end

Whole Flank

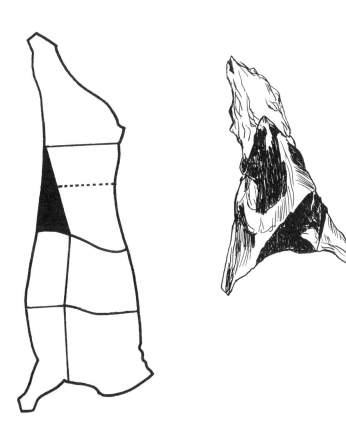

6. **Whole flank**—Aside from yielding larding fat and trimmings for chopped meat, this is the primal cut from which a true London broil (otherwise known as a flank steak) is cut. When of high-quality heavy USDA Choice grade, the London broil may be broiled. Otherwise it must be moist cooked because it is fibrous, lean, and sinewy.

7. **Round of beef** comes from the hind leg, a mobile section of the beef. It is the leanest cut of the hindquarter. The entire round usually weighs somewhere between 70 and 86 pounds before it is divided. Initially, a triangular cut known as the rump, which contains the pelvic bone, is separated from the top part of the whole round. The rump may be sold as is or boned out and rolled for moist cooking or, when of heavy quality Choice grade, for roasting. The remainder of the round is then further subdivided into six sections as indicated in the illustrations and described below.

Round

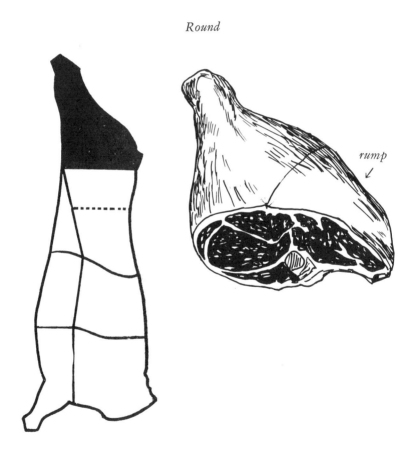

rump

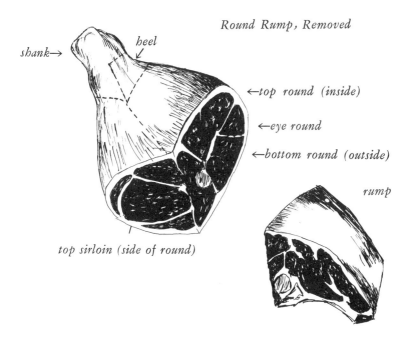

Round Rump, Removed

shank→   *heel*

←top round (inside)

←eye round

←bottom round (outside)

rump

top sirloin (side of round)

The side of round is also sometimes called either knuckle, veiny, top sirloin, or sirloin tip. It is the section of the round which is anatomically adjacent to the sirloin. The side of round is further subdivided into lean steaks and roasts of varying sizes and weights. One of the subdivisions of the side is the silver tip roast.

Top (or inside) **round** is the largest subsection of the whole round and the part most suitable for dry roasting. The whole top round is either halved or quartered into roasts or further cut into steaks.

Bottom round is cut from an exterior mobile muscle and is therefore best suited to moist cooking or tenderizing techniques such as marination. It may also be dry roasted with fat added to its exterior to keep it moist as it cooks. The whole bottom round is subdivided into steaks and roasts of varying weights and sizes.

Eye of round is a lean, oval, elongated muscle which may be further cut into roasts. The roasts may be dry roasted with added fat, or moist cooked.

The shank is cut from the shin section of the leg and is comprised of beef suitable for moist cooking and grinding.

The heel is located at the back of the lower part of the leg and yields meat suitable to moist-cooking methods and grinding.

Chapter 4

❖❖❖❖

The Beef Roast
from Coast-to-Coast

eight basic roasts

A roast is a cut of meat which is cooked uncovered in the dry, direct heat of an oven or on a rotisserie. Any beef you can roast can also be pot roasted or be cut into steaks for broiling or pan frying. If you have any doubts, ask the meatcutter or consult the glossary of beef cuts in Appendix 1 at the end of this book.

When you buy a roast there are several factors to consider. The general appearance of the roast is highly important, as is its flavor and juiciness. The most important factor, however, that factor that sets a beef tenderloin or rib roast ultimately above a round, sirloin, rump, or chuck, is *tenderness.* (Anyone who has ever tried to chew his way through a tough situation gets the point.)

The quality of tenderness is largely dependent upon the grade of beef you select, and from which primal cut of the carcass the roast was cut. These characteristics will determine just how much fibrous connective tissue, exterior fat, and interior marbling a beef roast will contain. The more connective tissue it has, the less tender it will be—the more marbling, the greater the tenderness. Although the exact cutting method and generic name applied to it may differ from one geographical area to another, the eight roasts described and illustrated here are basic roasts from coast-to-coast. They are listed in descending order of tenderness, and are USDA Choice grade beef:

1. Tenderloin roast
2. Strip loin roast
3. Rib roast
4. Top sirloin butt roast
5. Top round roast
6. Top sirloin roast
7. Chuck shoulder roast
8. Eye round roast

Whole Tenderloin roast

1. Tenderloin roast—The beef tenderloin, otherwise known as the filet mignon or Chateaubriand roast, is a boneless, slender, elongated cut from the loin. A small portion of this roast appears as the eye or filet section of the T-bone, Porterhouse, and sirloin steaks. The tenderloin can boast of the most tender meat of the entire beef carcass, and it is proportionately expensive—almost prohibitively so. The tenderloin is, therefore, most often not even sold as a roast but subdivided into individual-sized filet mignon steaks.

The tenderloin roast may be purchased whole in 4- to 6-pound sizes, or in halves of 2 to 3 pounds each. The narrowest end of the roast, which resembles a flap, is tucked underneath for cooking and the roast is tied and shaped. Extra fat may be added to its outside surface to compensate for the exterior fat that it lacks.

Beef tenderloin requires little cooking time and is always served rare. It may either be carved thickly into individual steak-sized portions or thinly sliced. There are about three servings to a pound.

Whole strip loin

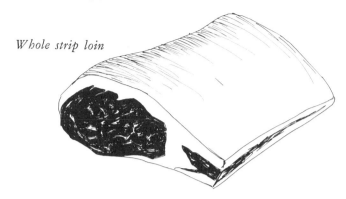

2. **Strip loin roast**—The strip loin roast is a boneless, practically wasteless roast from the top of the short loin. It is nearly as tender as the tenderloin roast, though slightly "grainier" in texture. Many consumers find its flavor superior to that of the tenderloin and it is equally as expensive. The strip loin roast is, therefore, most often subdivided into individual strip steaks.

The strip loin roast may be purchased whole at about 12 pounds or divided into halves or thirds. It is usually served rare, thickly carved or thinly sliced, depending upon preference. There are three or four servings to the pound.

3. **Rib roast**—The rib roast, which is cut from the rib of beef, is the only roast other than the tenderloin and strip loin that always guarantees tender, juicy slices throughout regardless of whether it is cooked rare, medium, or well done. The whole rib roast contains seven rib bones and is sometimes sold intact to restaurants specializing in buffets. More often, however, the whole rib is subdivided into three more practical, consumer-sized sections, each containing two or three ribs.

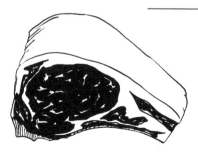

Rib roast first cut

First cut (or front cut) is the name applied to the first three ribs, those that are closest to the short loin. These are the only three ribs that really qualify as "prime ribs" (but should not be confused with USDA Prime grade quality). The first cut is priced slightly higher than the roasts cut farther away from the short loin on the rib. But it still constitutes a good buy, because it has less fat, less waste, and allows more carving leverage than the other roasts.

The first-cut rib roast can be purchased less expensively as is, untrimmed (bone-in), than semiboned. When it has been semiboned, the chine, which is the lower part of the back bone, and the plate or tail bone have been removed, and there are just two rib bones exposed in the center of the roast. This elegant, semiboned cut is what is commonly called a standing rib roast and is easier to carve than a bone-in roast, though slightly more expensive because of that additional trim. A boneless rib roast—also known as a Delmonico roast—has literally all of the bone removed.

Sometimes the first cuts are featured at a supermarket sale. If this is the case, they are most likely to be found with the bone in. Before you purchase a bone-in rib roast be sure that the tail or plate end has been cracked by the butcher. This makes the roast easy to carve after it has cooked. (In most supermarkets this practice goes without saying, but it never hurts to ask.)

The center cut is the two middle ribs and can be purchased like the first cut—either boneless, semi-boned, or bone in. It is just as tasty as the first cut, but less expensive because it contains more fat and waste. The center cut rib roast is easy to recognize. It has a large section of exposed fat between the third and fourth rib.

Rib roast center cut

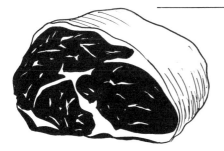

Boneless Delmonico roast

The end cut is the last two ribs and located closer to the chuck —a mobile part of the beef. It is, therefore, the least desirable of the rib roasts. These ribs weigh 3 to 4 pounds more than the other ribs because a large section of fat runs through the center of the cut, actually dividing the bottom part

Short ribs

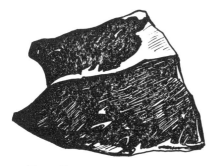

Top rib

*See page 23.

of the meat from the top.* The meat on top of the fat is more fibrous and closer in texture to the brisket. The meat beneath the fat is the same as the rest of the rib meat. So when the end cut is roasted and carved, the slices will not be uniformly tender. That is why it is a common practice among meat cutters to divide the end cut into three different sections: short ribs, top rib, and a Delmonico roast. When the end cuts are found on sale, make sure that the plate end has been cracked before you purchase it. This usually will have been done for you in a self-service market. (To make the most of this end-cut rib roast, see Chapter 20, "The Step-by-Step Guide to Cutting Meat.")

Whole top sirloin butt

4. **Top sirloin butt roast**—Also known as the hip, the top sirloin butt is comparable in price to the rib roast, but it is not as flavorful because it has less interior marbling than the rib. However, because the top sirloin butt is never sold bone in, it has less waste and more lean meat.

This roast is actually the center of the hip—an elongated 10- to 14-pound cut from the sirloin section of the steer. The cut is split in half for 5- to 6-pound roasts, quartered for 3- to 4-pound roasts, or further divided into steaks.

A top sirloin butt may be carved thickly or thinly, depending upon your preference. You may have to request this cut in advance because it is most often subdivided into consumer-sized steaks rather than roasts.

Whole top round   *Tied top round roast*

5. **Top round roast**—Of the six subdivided sections of the round primal cut—top, side, bottom, eye, shank, and heel—the top is the most desirable for roasting because it is inside and adjacent to the sirloin. Top round roast is a boneless, very lean cut that is in great demand because it contains very little fat and very little waste.

The roasts cut from the section of the top round nearest the sirloin are the most desirable, most uniformly shaped, and the most expensive. Cuts of roast farthest away from the sirloin are less uniformly shaped and more likely to be found on sale. The whole top round is usually what one sees cooked in delicatessens. Top round roasts generally weigh somewhere from 3 to 6 pounds and are best roasted with fat added to their exterior surfaces in order to prevent them from drying out as they cook. Like all very lean roasts, this roast should be carved into very thin slices for servings of maximum tenderness.

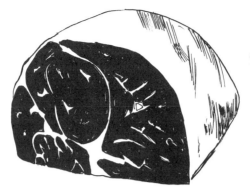

Top sirloin roast

6. **Top sirloin roast**—The top sirloin roast which is cut from the side of round is the second most desirable section of the round for dry roasting. This roast is usually sold in 3- to 4-pound sizes. One of the lean, less flavorful outside cuts of this roast is known as the silver tip.

Only a top sirloin roast of heavy USDA Choice grade—recognizable by a bright red color—takes well to dry roasting. Lower grades require moist cooking. Regardless of the section of the side from which it is cut, a top sirloin roast will always require extra fat such as cooking oil, shortening, or bacon added to its exterior surface if it is to be dry roasted. The top sirloin roast should also always be carved very thinly for maximum tenderness.

Chuck shoulder roast

7. **Chuck shoulder roast**—Only the first three or four pounds of the upper part of the chuck shoulder or cross rib—the part of the chuck that is closest to the rib—is suitable for dry roasting. And if you plan to roast it, it should be of high-quality USDA Choice grade. Extra fat added to the exterior surface of this very lean, boneless roast is a definite

asset. The chuck shoulder roasts cut farther away from the rib do not have the same quality and will require braising or pot roasting to make them tender. The chuck shoulder should always be sliced very thinly.

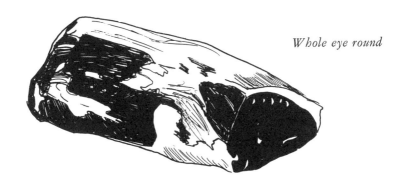

Whole eye round

8. **Eye round roast**—The eye of the round is the oval-shaped, small muscle of beef adjacent to the bottom round. It is a virtually wasteless, elongated roast which is about four inches in diameter. The whole eye weighs 6 to 8 pounds but is often divided into two roasts or further subdivided into steaks. This very lean cut should be ordered in advance as it is in very short supply. Eye round tends to be slightly dry when roasted, so choose high quality and add extra fat to its outside surface to help make it juicier when cooked.

Chapter 5

◊┼◊┼◊┼

All-American Steaks

family-sized... individual... skillet
...and one-minute steaks

It is probably safe to say that next to the economical, popular hamburger, a steak is an American's favorite meal. High-quality steaks are in great demand, and they are expensive. Yet nothing seems to satisfy the American appetite more than a tender, properly cooked beef steak, be it round, sirloin, strip, or Porterhouse.

One way to take the guesswork out of broiling large steaks is to purchase only those that are an inch or more thick. This particularly holds true if you prefer steaks rare to medium rare. Steaks less than an inch thick are apt to broil too quickly and become well done before you know it. These thinner steaks should be pan broiled or fried for greater assurance of tender, rare results.

Remember, too, the higher the quality of the steak, the faster you can expect to cook it. Most high-quality steaks have a liberal amount of marbling fat and marbling draws and holds heat.

Described and illustrated here are steak categories ranging from the family-sized, to the individual, the skillet, and the inexpensive minute-sized. They are listed in order of descending tenderness in each category and the generic names are given. All are conventional cuts available from coast-to-coast.

Because of supply and demand, the more tender, juicy steaks are more expensive. It isn't unlikely, however, to find your favorite steak on sale. Steak sales are quite prevalent because they draw customers into the store. (See Chapter 18, "Selecting Self-Service Meat.") By all means take advantage of the price cut, but be beware of a mark up on other products.

THE FAMILY-SIZED STEAKS

Here are the largest and most luscious steaks you can buy:

1. Porterhouse steak
2. Rib steak
3. Sirloin steak (hip)
4. Top round steak
5. Sirloin tip steak
6. Shoulder steak (chuck)
7. Chuck steak (blade)

Full loin rib end

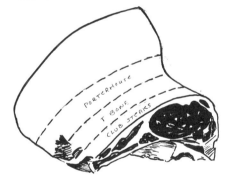

Short loin porterhouse end

1. **Porterhouse steak**—The Porterhouse is *the* superior steak, encompassing a large portion of the tenderloin or filet mignon and the strip shell steak. Only four Porterhouse steaks, each weighing 2½ to 3½ pounds can be cut from the short loin of beef. The short loin can, however, be further cut into T-bones (which are the same as Porterhouse except that they contain less of the filet mignon) and finally into club and shell steaks (which contain none of the filet mignon). See page 00.

Club

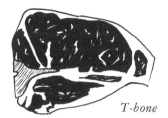

T-bone

Porterhouse, T-bone, and club steaks are equal in tenderness and can either be broiled or pan broiled. About 3 pounds of Porterhouse, T-bone, or club will feed a family of four.

Porterhouse

Rib steak

2. **Rib steak** is a fine-flavored steak and the finest steak available in Kosher markets. A subdivision of the rib of beef primal cut, this steak most often includes the rib bone, a liberal quantity of exterior surface fat, and is well marbled.

Two bone-in rib steaks should be adequate for four people. When the rib steak is trimmed and boned it is sold as an individual steak called Delmonico or rib-eye. The rib, Delmonico, or rib-eye may be broiled if thick, or pan broiled if cut more thinly.

3. **Sirloin steak** (*hip*) is cut from the end of the full loin or hip and is a favorite with consumers because it actually has less waste and more lean meat than the Porterhouse. The section of the primal cut from

which a consumer-sized sirloin is cut will determine its size, shape, and bony structure. Pictured here are three sirloins which contain different sections of bone. They are:

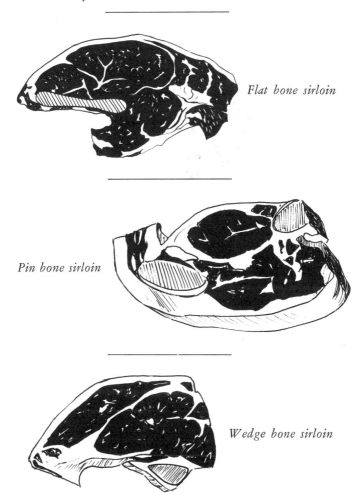

Flat bone sirloin

Pin bone sirloin

Wedge bone sirloin

Of the three, the flat bone is the most desirable—the pin-bone is closest to the short loin and contains a good portion of the tenderloin, but also a lot of bone and is the most wasteful sirloin steak. The wedge bone is the least wasteful but also the least tender, being closest to the round. Those sirloin steaks cut farther back on the hip—closer to the rump end of the round—are usually boned out and offered to the consumer under various names such as shell hip, boneless sirloin, or top sirloin butt steak. A 3-pound bone-in or a 2-pound boneless sirloin will

Boneless sirloin

serve four people generously. This steak may either be broiled or pan broiled.

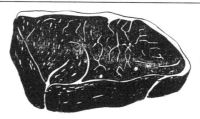

Top round steak

4. **Top round steak**—which comes from the cut that is also known as inside round—makes a fine flavored steak. It is a lean cut that has very little waste and no bone. It is not, however, as juicy and tender as the Porterhouse, rib, and sirloin steaks.

Because top round steak is so lean and has little marbling, it is a good steak for weight-conscious consumers and is in great demand. One and a half pounds serves four people. The steak may be broiled with a little additional fat added to its exterior surface, but it will be even better if it is pan broiled or pan fried.

Sirloin tip steak

5. **Sirloin tip steak**, which comes from the side section of the round, is also known as round, side of round, knuckle, veiny, or top sirloin. It should not be confused with regular sirloin which is only cut from the hip end of the full loin.

The highly popular sirloin tip is less tender than regular sirloin, but has the virtue of having less waste. It is a bright red, lightly marbled, oval cut of beef about six to eight inches in diameter. One and a half pounds will feed a family of four—or allow 6 ounces per portion. Recommended cookings methods: pan broiling, frying, sautéing.

Shoulder steak

6. **Shoulder steak** (*chuck*), which is cut from the chuck, is the second-best selection available in a Kosher market. It is a low-waste, fine-flavored cut. It is also the most uniformly shaped steak that can be cut from the chuck—rather elongated and slightly smaller than the sirloin tip. The shoulder steak is delicious either pan broiled or fried. One and a half to two pounds feeds four hungry people.

Chuck blade steak   *First-cut chuck steak*

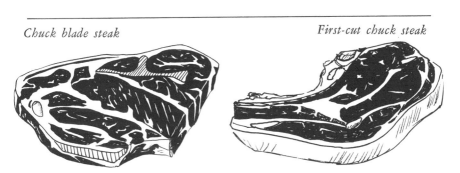

7. **Chuck steak** (*blade*) of USDA heavy Choice is frequently on sale as a special. As such, it is a good buy because it is tender and juicy. The first three or four steaks cut from the section of the chuck closest to the rib are particularly desirable. It is possible to broil these steaks, although braising is a better overall method of cooking. A 3½-pound chuck steak, which contains a good deal of waste, both the blade bone and some cartilage, serves four people.

INDIVIDUAL STEAKS

These steaks are well-trimmed, single-sized portions of the family-sized steaks listed earlier. The filet mignon, for example, is the most tender section of the Porterhouse and sirloin steaks. All of these cuts can be differentiated from the skillet and minute steaks in that they can be broiled. They are listed here in descending order of tenderness:

1. Filet mignon steak
2. Strip loin steak
3. Delmonico steak
4. Rib-eye steak
5. Top butt sirloin steak
6. Chuck tender steak
7. Top round steak

Filet mignon

1. **Filet mignon steak** is a famous, very elegant cut, and it is also very expensive. The filet mignon, which comes from the whole beef tenderloin, is the most tender section of the Porterhouse and sirloin and it can also be found to a small degree as part of the T-bone and sirloin. The high cost of the filet mignon can be accounted for in the trimming. Nearly a pound of meat is required to yield a 6- to 8-ounce portion—or enough to feed one person.

2. **Strip loin steak**—also referred to as strip steak—is a boneless cut from the short loin. Untrimmed, with the bone intact, the strip loin is known as the T-bone, shell, New York or club steak. Depending on the thickness to which the strip is cut, it will contain more meat than the filet mignon, generally up to 10 to 16 ounces.

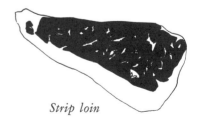

Strip loin

Delmonico steak

3. **Delmonico steak** is sometimes called "Spencer steak" and is a cut trimmed from the rib-eye of beef. It is usually found in individual-sized portions of 8 to 10 ounces.

4. **Rib-eye steak** is also cut from the rib of beef and weighs 4 to 8 ounces. There is only one difference between the rib-eye and the Delmonico: the rib-eye is totally trimmed and devoid of fat and waste.

Ribeye steak

Top butt sirloin

5. **Top butt sirloin steak** is sometimes called filet of sirloin or shelled hip. It is a boneless cut located close to the rump section of the round. This is a special cut usually found in 6- to 8-ounce individual-sized portions.

6. **Chuck tender steaks** are small cuts that look very much like the rib-eye steaks—but they are the eye of the chuck and not as well-marbled or tender. Only USDA heavy Choice cuts from the section of the chuck nearest the rib are, in fact, suitable for broiling. The chuck tender is usually found in 4- to 6-ounce individual-sized servings.

Chuck tender steak

Top round steak

7. **Top round steak** of USDA heavy Choice grade is an attractive purchase for the diet-conscious beef lover. Also known as the inside round, it contains no waste and is totally lean meat. For individual-sized portions you may purchase a whole 1½- to 2½-pound top round steak and simply divide it into four portions. Before broiling, add a little fat or oil to the surface of this cut because it contains very little marbling.

SKILLET STEAKS

These skillet steaks are small, boneless beef cuts of USDA Choice grade. They are only about half the thickness of the individual steaks and are best adapted to pan frying. For best results and maximum tenderness these steaks will need to be cooked in a little fat or oil.

1. Top round steak
2. Top sirloin steak
3. Shoulder steak
4. Chicken steak
5. Bottom round steak
6. Eye round steak
7. Flank steak

Top round steak

1. **Top round steak,** cut from the inside of the round, is a meaty, all-lean cut. It is usually sold in thin, 4- to 6-ounce slices—slices so thin they cook almost instantly. Top round makes excellent steak sandwiches.

Top sirloin steak

2. **Top sirloin steak**, also known as knuckle, veiny, or sirloin tip, comes from the side of the round and is nearly comparable to the top round for the qualities of tenderness and juiciness. Top sirloin is cut into a thin, oval steak which can be further divided into quarters at home for easy cooking. A 1½-pound cut will make about four individual-sized servings.

Shoulder steak

3. **Shoulder steak** (or chuck) is usually a good buy because it is less expensive and almost as tender as the top sirloin. The shoulder steaks are cut from the chuck and may be purchased in sizes ranging from ¾ to 1½ pounds. One and a half pounds will provide enough meat to be cut into four individual servings.

Chicken steak

4. **Chicken steaks** cut from the chuck are called among other things: "Kalickle," "petite steak," and "club" (chuck section). In fact, these chuck steaks probably parade under more aliases than any other cut of beef. They are usually sold in individual-sized portions of 4 to 6 ounces.

Bottom round steak

5. **Bottom round steaks** are cut from the outside of the round and are lean and flavorful. If not of USDA heavy Choice grade, however, they tend to be quite tough. Therefore, it may best ensure tenderness to purchase this cut cubed. Cubing, which is a machine process, breaks down the fibrous inner connective tissue of the beef and makes it tender. One and a half pounds of bottom round will provide four ample servings.

Eye round steak

6. **Eye round steaks,** which are small, disc-shaped, 3- to 4-ounce slices from the muscle adjacent to the bottom round section, tend to be somewhat dry. That is why, when not of USDA top quality Choice grade, this steak is usually found cubed.

Flank steak

7. **Flank steak** is the true London Broil—a tasty, lean cut of beef which weighs about 2 pounds. To qualify as a skillet steak, the whole flank may be cut horizontally into ten to fifteen slices and then pan broiled or pan fried.

THE ONE-MINUTE STEAKS

A real value for consumers, these individual-sized portions of boneless, small, thinly cut steak differ from those described previously in that they are only about half as thick, less expensive, designed for expediency in cooking, and always pan fried or sautéed. These steaks will cook instantly. For best results a little shortening should be added to the pan when you cook them, and the pan should be preheated. After the steaks are added to the pan, they should literally be cooked but thirty seconds per side—one minute totally.

1. Top round steak
2. Top sirloin steak
3. Chicken steaks
4. Eye of chuck steak
5. Bottom round steak
6. Shoulder steak
7. Flank steak

1. **Top round steak** is cut from the inside round and is extra lean.

2. **Top sirloin steak** from the side of round is very tender, has a good flavor and fine texture.

3. **Chicken steaks,** which are also called petite steaks, kalickle, or club (chuck section), are a juicy triangular cut which contains a section of gristle in the center. The gristle is usually trimmed away—leaving a split in the center of the steak.

4. **Chuck tender steak** is a continuation of the rib-eye muscle in the chuck section. It has a greater degree of fat than the other chuck steaks, and it is tender and juicy.

5. **Bottom round steak** cut from the outside, of USDA heavy Choice grade, has quite a bit of elastic connective tissue, but is nevertheless quite tasty. This steak may be braised as well as pan fried.

6. **Shoulder steak** cut from the end of the chuck is sold either very thinly sliced or cubed. It has a robust flavor and a well-developed texture.

7. **Flank steaks** are tasty steaks cut from the tail end of the loin of beef. The extra-long tail on a T-bone is part of the flank meat.

Chapter 6

❂❂❂❂❂

The Moist-Cooked Cuts

beef for pot roasting, braising, stewing, boiling…and the "London broils"

Beef cuts cooked by one of the "moist methods" are of the less tender and less expensive variety. Moist cooking is a slow, low-temperature affair. Whether you plan to pot roast, braise, stew, or boil a piece of beef, you will require a certain amount of liquid, a tight lidded pot, and plenty of time.

Braising and pot roasting are virtually the same method of meat cookery. There is only one difference. Pot roasting requires more time. The beef cuts used are simply larger and chunkier, weighing from 3 to 5 pounds or more. Those cuts described in this chapter to be used for braising weigh about 1½ to 3 pounds.

Otherwise, the same basic cooking procedure applies to both pot roasting and braising. It begins with browning the beef on all sides in a little hot shortening, flouring it first if a richer, more robust flavor is desired. A heavy cast-iron pot works best—but any pot with a tight-fitting lid will do if the heat is kept sufficiently low. After browning, the beef is seasoned according to taste or to recipe. Then hot liquid is added gingerly—one half cup or less should do the trick. Then the pot is closely covered. The meat should remain just at the simmering point throughout the cooking process. This can either be accomplished on top of the stove or in a moderately slow oven (325°F.). If the liquid cooks away, some more can always be added.

Depending on size and cut, a pot roast may require upwards of 3½ hours to cook, and a stew and a braised steak as little as one hour. It is impossible to calculate the time exactly—but the pot should be watched and the end results should be fork tender and tasty if a little culinary art has been applied. Some shrinkage is normal when beef is moist cooked, but overcooking will cause beef to shrink to a greater degree and lose all of its natural moisture.

BEEF CUTS TO BRAISE

There is a wide variety of beef you can braise. Many meats adapt successfully to this method of cooking because the slow cooking process creates steam inside the covered pot and that softens the tissues and makes the meat tender. Visually, many of the braising cuts of beef resemble a double thick conventional steak, and about a pound will be required to feed three to four people or more depending upon how it is carved. Recommended below in descending order of tenderness and leanness are seven such cuts available everywhere:

1. Top round
2. Top sirloin
3. Chuck shoulder
4. Bottom round (with eye)
5. Chuck blade steak
6. Flank steak
7. Short ribs

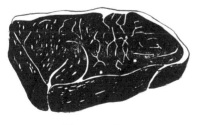

Top round

1. **Top round** has a grainy texture that, when cut to a thickness of 1½ to 2½ inches, adapts beautifully to braising. Top round is always lean, has a fine flavor, and because it has some marbling and less connective tissue requires less time to cook to the tenderness point than the other cuts listed below.

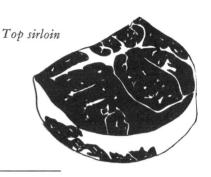

Top sirloin

2. **Top sirloin** cut from the round is an elegant, oval, boneless cut that is just as tasty as top round but is less tender and so will require more time and expertise to make it tender.

Chuck arm

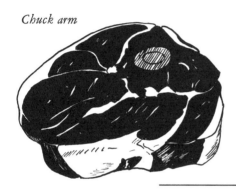

3. **Chuck arm steak** is also known as shoulder or arm bone roast. It is a lean, tender, and uniformly shaped cut that is ideal for braising because it will cook very evenly throughout. Chuck shoulder is cut into thick 2- to 3-pound steaks for braising.

Bottom round with eye

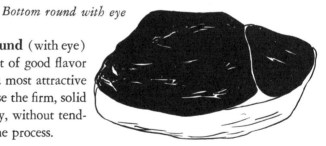

4. **Bottom round** (with eye) is an elongated cut of good flavor and texture. It is a most attractive cut to serve because the firm, solid meat cooks quickly, without tending to shrink in the process.

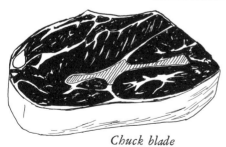

5. **Chuck blade steak** offers meat that is juicy and well flavored but not as lean as bottom round (with eye). The blade roast braises more quickly, however, due to the bone it contains, which conducts the heat.

Chuck blade

Flank

6. **Flank steak** or London broil (flank) is the only beef for braising that doesn't require shaping by a butcher, but is ready for the pot just as it is cut from the steer. Only a thin, outer layer of membrane is trimmed away. For best results, carve this lean, delicately flavored cut at an angle on the bias.

Short ribs

7. **Short ribs of beef**—also known as English ribs—might top this list for flavor, but they are in great demand and often hard to come by. Short ribs contain the ends of the rib bones, so up to 3 pounds may be required to supply four hearty servings.

POT ROASTS

Pot roasting is a cooking method that will make most anything tender—but the most demanding and discriminating cooks will settle for nothing less than USDA Choice grade for their pot roasts. Why? Because a USDA Choice grade pot roast guarantees both tenderness as well as a robust flavor essential to this hearty meal. However, undergraded beef such as USDA Good may be used also. The basic pot roasts listed below are in descending order of tenderness and flavor. All are available country-wide. A pound of boneless pot roast will serve three or four or more depending upon shrinkage and how it is carved.

1. Center-cut chuck
2. Chuck shoulder
3. Rump roast

4. Bottom round (with/without eye)
5. Top rib
6. Brisket
7. Plate

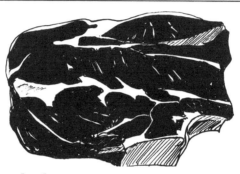

Center-cut chuck

1. **Center-cut chuck** is cut from the center or nearer to the rib end and makes a most delicious and attractive pot roast. The meat has a good texture and is also lean. Available in cuts from 3 to 5 pounds, this pot roast can be puchased as cut from the carcass or de-boned, then rolled and tied.

Chuck shoulder

2. **Chuck shoulder** from the upper part of the chuck makes a very shapely pot roast. The shoulder contains only one small bone and a flat piece of gristle which may be removed by the butcher. The result is a very lean, easy-to-carve, uniformly shaped pot roast. The flavor, however, is not quite as distinctive as that of center-cut chuck.

3. **Rump roast** is a triangular cut from the tail section of the round. It can be had with the pelvic bone intact in 4- to 6-pound sizes, or

Standing rump roast

de-boned, rolled, and tied. With the bones it is known as a standing rump roast. The flavor is good, but the meat contains a good deal of connective tissue and a sparse outer covering of fat.

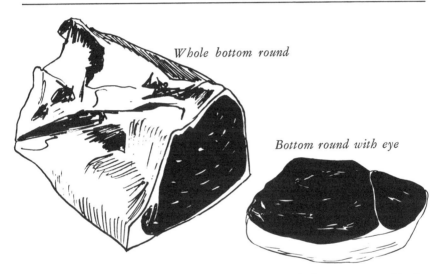

Whole bottom round

Bottom round with eye

4. **Bottom round** is lean and easy to identify by its unique blocky shape. The most select portions of the carcass from which to cut these 3- to 4-pound sections are the center and the rump end of the round. When bottom round is cut into a roast in such a way as to include the eye of round it makes an extremely popular pot roast which is thus quite scarce. It must, therefore, often be ordered in advance. The bottom round (with eye) is available in both 3- to 4- and 6- to 8-pound cuts.

5. **Top rib** provides a flat 3- to 4-pound pot roast. This is the cut that is taken from the top section of the last two beef ribs. Visually and texturally it closely resembles a straight cut of brisket. The meat may be

Top rib

easily carved into long, wide slices that provide a rich, robust flavor. For best results, the top rib should be thoroughly trimmed of fat.

Brisket

6. **Brisket** is a solid, well-formed and distinctive cut from the fore-quarter of beef. This tricky cut contains a good deal of connective tissue and a great deal of surface fat and can have a lot of waste. The least wasteful, best brisket buys come from the thin end and are usually labeled "straight cut," "flat cut," or "square cut." Less desirable and more waste-ful brisket is cut from the thick end or back end. At double the price for equal weight, the cut from the thin end would still be a more economical purchase. There will simply be more meat and less waste. Brisket makes an extremely tasty pot roast, but it is best to insist on a well-trimmed piece, and long, slow cooking is essential. Cured, cooked brisket is corned beef.

---

7. **Plate** meat, with or without its bone, makes a fine pot roast. The slices are not quite as uniform as those of brisket, although the cut itself is very similar. This cut is inexpensive and tasty, but does contain a great

Plate

deal of waste. Three or four bone-in cuts should suffice for the same number of people.

BEEF FOR STEWING

Stews can be a very tasty way to economize. A stew is what you do to a less tender cut of meat. Depending upon the liquid and flavoring used in stewing, even undergraded beef (such as USDA Good) will make a perfectly acceptable, even delicious stew.

To make a stew, follow your recipe or your inclination. The basic method is very simple. You begin by browning the meat on all sides in hot fat—flouring it beforehand if you prefer a deeper brown. Then you add several cups of hot liquid: stock, bouillon, wine, water, or a combination (the Belgians made a great stew with beer). Season according to your tastes. Cover the pot closely and cook slowly taking care that the meat is *simmered* until tender, and not boiled. Vegetables may be added near the end of the cooking time. The time will vary according to the cut of meat used. Meat is done when it yields easily to the points of a fork. When everything is just tender—not overcooked—the gravy may be thickened if necessary. You are ready to eat.

The most tender and flavorful stews can be achieved if the beef you choose is solid and has a little marbling but not too much gristle or fat. Meats will shrink considerably during the stewing process, so make sure the cubes are not cut too small. Beef cut in 1½- to 2½-inch cubes makes an attractive dish. A quarter pound should suffice per serving combined with other ingredients.

Almost any beef can be cut into cubes and stewed. The following beef cuts, however, are listed from one to seven in descending order of their ultimate adaptability to yield solid, flavorful results in a stew:

1. Boneless chuck (center)
2. Boneless chuck (inside or rib)
3. Heel of round
4. Flank steak
5. Top rib
6. Shin of beef (hindquarter)
7. Plate

Boneless chuck (center)

1. **Boneless chuck** (*center*) has a good texture and is generally lean throughout. The chuck texture lends richness to the stew and holds up well during the long, slow cooking process. A bone-in chuck steak may be cut into cubes for stewing at home.

2. **Boneless chuck** (*inside or rib*) is a square cut from a delicate section of the chuck which is often cut and sold labeled as London broil—in 1½- to 3½-pound sizes. Cut into cubes for stewing, it is succulent, rich with flavor, and ultimately very tender. This chuck is equally as good for stewing as chuck center, but in short supply. Thus it is often necessary to order this cut in advance.

Boneless chuck—(inside)

Heel of round

3. **Heel of round** is cut from the extreme end of the round. It is even more tender than chuck, but does not have the fine flavor. Because it is extremely lean, USDA Choice grade is best for stewing. Two pounds of heel of round stewed with other ingredients should be ample for four servings.

Flank

4. **Flank steak** has an unusual flavor when braised and makes an equally good stew. Because it is thinner, it requires less cooking time than the other cuts mentioned. One flank steak should make enough stew for a family of four. If it is purchased whole, cut the bulk of the steak into uniform-sized cubes for stewing, then add the uneven ends to the stew pot for extra flavor.

Top rib

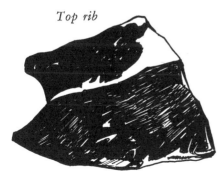

5. **Top rib** is the tender, flavorful upper portion of the last two ribs of the beef. It is often sold rolled, but can also be cut into uniform cubes for stewing. Two pounds should be sufficient for four.

6. **Shin of beef** (*hindquarter*) is a very low-cost cut. But because it contains an elongated tendon, it will require longer cooking. In fact, shin meat from the forequarter is too fibrous to even adapt well to stewing. One must, therefore, specify when purchasing shin beef that it be cut from the hindquarter. This cut costs less than the others listed, and is tasty and lean.

Shin

7. **Plate** is an inexpensive beef that is usually sold with the bone intact. It may have to be cut into cubes for stewing at home. Plate meat has a delightfully rich flavor, but it does contain a good deal of waste.

Plate

SEVEN BASIC BEEF CUTS FOR BOILING

Boiled beef is more accurately described as "simmered beef." Although some recipes require that beef be plunged directly into rapidly boiling water for five minutes in order to firm its texture and seal in its juices, as a rule the cooking water should never stay at a rolling boil, but merely stir on the surface in a gentle simmer.

Meats prepared according to this ancient and most economical cooking method should be well trimmed of excess fat prior to cooking if possible. Then, unlike those meats that are stewed, these are entirely immersed in liquid.

Timing is of the essence here. Beef should be boiled only until just tender all the way through. Even results are achieved by selecting uniform cuts. Beef that is underboiled will be tough, chewy, and unpalatable. But beef that is overboiled is bound to be bland, shred, fall apart—and all of its nutrition will be lost in the broth.

The least tender beef cuts, those with the most tendons and ligaments, lend themselves beautifully to boiling. More tender cuts would become stringy cooked in this manner, but those seven described below should remain firm with careful timing. They are recommended in order of flavor and tenderness—and most are a real boon for budgeters:

1. Brisket
2. Short ribs
3. Heel of round
4. Neck-end chuck
5. Shin meat (hindquarter)
6. Plate
7. Shin meat (forequarter)

Brisket

1. **Brisket** from either the thin or thick end will make a tasty boiled meal. The thick-end cut will have nearly twice the waste, however, so unless it is marked at half the price, purchase the thin end. Brisket is characteristically a solid, fibrous cut, which makes it perfect for boiling. It is also easy to carve and yields uniform slices that have a highly distinctive flavor. One quarter to ⅓ pound will suffice per serving.

2. **Short ribs** of beef, known as "flanken" in Kosher markets, are closely related to the brisket. They are cut from approximately the same anatomical area. Like the brisket, short ribs are rich with unusual flavor. However, they are in somewhat short supply both

Short ribs

boneless and bone in. With the bone you will need nearly 3 pounds for four servings.

Heel of round

3. **Heel of round** has less flavor than the brisket, but it also has less waste and more lean meat per pound. This cut from the extreme end of the round contains a great deal of connective tissue which makes it ideal for boiling. Purchase ¼ pound of boneless heel per serving.

4. **Neck-end chuck** purchased with or without the bone is an extremely fibrous cut. Of all the cuts recommended for boiling, it is the leanest. Neck is also somewhat irregularly shaped, which makes for irregular carving. You could possibly end up with "chunks of chuck" rather than slices. The flavor, however, is exceptionally good. One quarter pound of boneless chuck neck per serving.

Chuck neck

Shin (hindshank)

5. **Shin meat** cut from the *hindquarter,* contains a good, flavorful marrow bone, but may also be prepared boneless. Shin meat contains tendons and ligaments, but boils up juicy and tender. It will serve most admirably as the basis of a highly economical meal. One quarter pound of boneless shin provides a serving.

6. **Plate beef** adapts most tenderly to simmering and has a hearty flavor. Look for it in one of two economy cuts: thin-end plate with bones that resemble short ribs, or thick-end plate, which looks twice as large but contains a lot of inner fat and waste. Purchase three or four bone-in cuts for the same number of people.

Plate

Shin (foreshank)

7. **Shin meat** from the *fore-quarter* has the same flavor as shin meat from the hindquarter, but the tendons and ligaments are more well developed and the exterior membrane is thicker, so it requires considerably longer cooking time in order to make it tender. One quarter pound of boneless shin is equal to a serving.

SEVEN "LONDON BROILS"

No chapter on moist-cooked cuts of beef would be complete without a footnote on those that adapt best to London broil cookery. Essentially, London broil cookery is braising. Flank steak, the only true London broil, has been discussed previously under those meats suitable for braising.

In markets today, however, many areas of the beef carcass are cut and labeled London broil. In many states the law now requires that the primal cut also appear on the label, in which case the label reads like this: London broil (chuck). Listed here in order of tenderness are seven such cuts one is likely to find. You will need a quarter pound of these boneless cuts per serving. They all adapt well to braising, and some can even be broiled with delicious results:

1. London broil (sirloin)
2. London broil (top round)

3. London broil (side of round)
4. London broil (chuck shoulder)
5. London broil (bottom round)
6. London broil (first-cut chuck)
7. London broil (chuck)

1. **London broil (sirloin)** is in reality a sirloin steak cut double thick with the bone removed. If well marbled, it may be broiled, but if it is very lean or undergraded, it will be more tender braised. It is usually available in sizes of about 3 to 4 pounds.

2. **London broil (top round)** is a very lean yet tender cut. It may be purchased in sizes ranging from 1½ to 2½ pounds.

3. **London broil (side of round)** is a section of the whole round and equally as lean, though not quite as tender as top round. This cut is also known as top sirloin.

4. **London broil (chuck shoulder)** is similar in quality to the cut from the side of round described above, but not quite as oval in shape. Select this tasty, lean cut from the center slices if possible. The end cuts of shoulder have more connective tissue, and when that tissue is removed the meat is split.

5. **London broil (bottom round)** is very lean but the least tender of the round cuts for London broiling. It will therefore require longer braising.

6. **London broil (first-cut chuck)** has an excellent flavor. The first cuts have only a long, thin strip of cartilage, which becomes bone as the chuck is cut farther down.

7. **London broil (chuck)** is sometimes called kalickle, chicken steak, chuck tender, or when it is deboned for the London broil method of cooking, filet of chuck. The result is flavorful and tender.

✿✿✿✿

Chapter 7

Chopped Beef—
Everyone's Favorite

Americans consume 40 billion hamburgers a year according to one recent estimate. If a pound of ground beef can be made into four hamburger patties, that means that 10 billion pounds of ground beef per annum is gobbled up just for hamburger. And that doesn't include the beef ground for meat balls, meat loaf, meat sauces, various casseroles, and steak tartare. That's a lot more ground beef. In fact, one can safely conclude that chopped beef is a U.S. household staple—and with good reason. It is versatile, reasonably priced for the most part, and more importantly—delicious.

To qualify for the meat counter, commercially ground beef cannot contain any seasoning, additives, or artificial coloring. And the ratio of fat to lean must not exceed a 30 per cent maximum. That's the amount permissible by law. An ideal hamburger, however, should have a fat content of only 20 to 25 per cent. At least 20 per cent fat is desirable because it adds flavor and juiciness to ground beef like marbling does to solid beef. (A pinkish color indicates an inordinately high amount of fat.)

Ground chopped beef should be used as soon as possible after it is purchased, because bacteria growth rapidly accelerates after meat is chopped. But it can be stored for a day or two in the coldest part of the refrigerator. If you are not planning to use the ground beef you bought on Monday morning by Wednesday night, it should be frozen.

Ground beef also has a tendency to discolor if it isn't cooked right away. Sometimes patches of the beef remain red while others appear a grayish brown. Two reasons for this are that the butcher ground a mixture of older beef trimmings with fresh-cut beef, or he mixed fresh beef with frozen. At any rate, unless the discolored meat has a distinctly gamey odor, it is usable although certainly not as desirable.

Recipes such as steak tartare call for extra-lean, twice-ground beef. Grinding beef a second time breaks down, smoothes, and tightens the texture of the meat, and thoroughly mixes the fat with the lean. Coarsely ground beef is best for hamburgers—beef that has only been sent through the grinder once. The lean and the fat are loosely packed so the patties will be juicier and not as firm in texture.

Which brings us to the question: which beef is best to grind? Most supermarkets offer ground beef from ground round, ground top sirloin, or ground chuck. Actually the best hamburger flavor is the economically priced ground chuck. Ground chuck has a higher fat content than ground round which makes for more juicy burgers, and it costs 20 to 30 cents less than ground round.

If a butcher is doing the grinding before your eyes, select the center chuck and ask him to combine it with chuck from the neck end. The chuck from the outside shoulder may also be used but it has less flavor and contains more moisture which tends to cook away. The prewrapped ground chuck you are likely to find in the supermarket meat counter is probably a blend of these three. Lean chuck is best for making meat loaf, too. It holds its shape firmly and combines well with ground lamb, veal, and pork.

If you're dieting, you may want to choose another variety of ground meat. Here are some of the alternatives:

Ground top round is more costly but is far leaner than ground chuck. However, it lacks chuck's fine flavor. Pure ground top round is something of a rarity because most of the top round is used for roasts and steaks. In fact, only 10 per cent of the total round is ultimately ground. That ground in the supermarkets may contain a small percentage of top round trimmings, but it is more likely to be parts of the heel. The best way to make sure you are getting pure ground top round is to request that it be ground for you.

Ground top sirloin comes from the sides of the round or the ends of the sirloin steak. It is lean and more flavorful than round, and sometimes is even priced lower.

Ground heel of round is very economical. It has a deep red color and is extra lean, but it lacks the flavor of both ground top sirloin, top round, and chuck. If more flavor is desired, the ground heel of round can be successfully combined with another more fatty kind of chopped beef.

Ground flank steak is not usually displayed ground, so the consumer doesn't generally think of it when purchasing chopped meat. He should. Ground flank steak has a unique and mildly delicious flavor that isn't inherent in any of the other kinds of chopped beef. The smaller the flank steak, the leaner its meat will be. A 1½- to 2-pound flank yields meat of exceptional leanness.

Chuck and the other ground meats mentioned above can also be combined with lower-cost cuts and save you a great deal of money. Ground beef shin, for instance, combines well with ground chuck. The chuck lends flavor and texture to the shin of beef, the ground shin laces the fatty chuck with lean. For hamburgers, ground shin beef by itself would be a dark, dense disaster, but mixed with chuck it is tasty and low cost too. Ground flank trimmings also combine most economically with ground heel of round.

There are, in fact, almost infinite ground meat combinations. If you are addicted to chopped beef, the best advice is to experiment, especially when planning to buy in quantity for a family freezer. Try various blends in small quantities until you find the one that best suits your taste and your pocketbook.

Some popular roadside eateries that specialize in hamburgers have discovered the secret of making eight hamburger patties from a pound of beef and providing their customers with 2-ounce portions on plenty of bun. For home consumption, a pound of chopped beef will yield anywhere from two to four hamburgers, depending upon the amount of fat, their size, your appetite, and how you plan to eat it. To get four perfect 4-ounce patties from a pound of ground beef, simply halve the pound, then halve the half. Or divide each 4-ounce hamburger into quarters for 1-ounce meat balls. Instant portion control! The trick is to handle the meat as little as possible so that it does not get too compacted and lose its natural juiciness. To prevent the meat from sticking to your hands when you are forming the patties, first run your hands under cold water. If you are planning to freeze the portions for any length of time, omit the flavorings and most especially salt. Salt prevents the ground beef from freezing rapidly and therefore reduces the time it may be frozen. For this reason, Kosher chopped beef is not recommended for freezing because it has been washed in a saline solution.

This discussion of chopped beef and hamburgers would hardly be complete without mentioning those innocuous processed "beef patties" or "meat patties" one often finds in the supermarket freezer section. The typical patty of this type contains but a percentage of beef. You, the consumer, are also paying for pork spleens, soya flour, salt, dextrose, hydrolized vegetable protein, flavoring, and water. In fact, by law "beef patties" cannot be called "hamburger" simply because they are not pure beef. Yet most processed patties are more expensive than those you can easily make yourself with pure beef.

In this age of instants and convenience, the hamburger is one of the very simplest dishes to prepare. It makes common sense to take advantage of sales, especially when purchasing large quantities for the freezer. Another budget booster: lean ground meat (such as sirloin, round, and flank) combined with other ground cooked meat stretches leftovers into a meat loaf and another meal.

Chapter 8

꘠꘡꘠꘡꘠꘡

Them Bones and the Gold Mine in Your Garbage Pail: Stock

definition… techniques… ingredients

There's a gold mine in your garbage pail that you've been overlooking if you're in the habit of tossing out scraps of used meat and bone. These are the basic ingredients of stock. And good stock can be the single ingredient that turns an ordinary dish into haute cuisine. Use stock rather than water in the next stew you prepare and taste the difference. Almost any recipe, in fact, that calls for a cup of water or a cup of stock will be better by far if the stock is used. That goes for gravies, sauces, soups, aspics, and marinades.

Stock is the rich juice extracted from meat and bones by slow simmering. It is solid meat flavor, protein, enzymes, vitamins, fats, mineral salts, and gelatin reduced to liquid form. It is the very essence of meat. Strained, degreased, and clarified beef stock becomes beef bouillon. Reduced by further cooking beef bouillon becomes consommé. Reduced even further it's double consommé—and finally it will become meat glaze. The canned commercial beef bouillon and consommé are simply no substitute for the real thing. Making your own stock can be a very rewarding experience, and a very thrifty habit, too.

Stock does not take a great deal of skill to make, but it does require a certain amount of patience. Anywhere from two to twelve hours' worth —according to which cookbook you read. Some experts prefer the old-fashioned methods and simmer their stockpots overnight, while other more modern practitioners of stock cookery claim that excessive cooking destroys food value and produces a bitter flavor. Whichever theory you adhere to will also require a large heavy pot, cold water, a few vegetables (and/or vegetable parings such as celery leaves, carrot and leek tops, and parsley stems), peppercorns, herbs, spices, meat, and bones. Some recipes call for salt and others omit it because stocks are seldom used in their original form. The exact stock recipe may very well depend, in fact, on what you have on hand. The scraps from Sunday's roast dinner plus a couple of other bones you'd stashed away in the freezer are all the incentive you should need to start a stockpot. Make a large quantity, refrigerate it overnight, then next day de-fat it, pour the stock into airtight containers and freeze to always have stock on hand for your favorite recipes. It should keep well without losing flavor for up to six months in the freezer. It must, however, be immediately reboiled after it has been defrosted.

Bones are essential to stock, but too many of them in proportion to the meat can spoil the broth by making it gluey and suitable only for sauces that need thickening. The ideal stockpot will contain about 50 per cent meat to 50 per cent bone. Bones, whether fresh or cooked, have a tendency to spoil quickly. If you are not planning to use them immediately, it is best to stick them in the freezer. And, if purchasing bones specifically for stock, your investment need not ordinarily exceed $1.00.

Basically, meat stocks fall into two categories: white and brown. Veal meat and bones (and light-colored vegetables) form the basis of white stock; beef and its bones the basis of brown. Pork, ham, and lamb are so distinctively flavored that their stocks are called for only in recipes that capitalize on their highly unique qualities of flavor. Split pea soup, for instance, calls for ham bone, and Scotch broth just wouldn't be without its lamb bone.

A hearty brown stock requires meat with lots of connective tissue which contains the necessary gelatin. It also requires marrow bones—preferably cut into small pieces to expose more surface areas to the liquid and thus enable a greater degree of flavor and nutritional extraction. Many recipes also require that the meat and bones be roasted initially in order to bring a heartier flavor to the stockpot.

The exact proportion of solid ingredients to water varies from recipe to recipe. If starting from scratch, and you have never made stock before, consult your favorite cookbook before embarking. Listed below are some suggested cuts for brown stock. Veal bones also have a good deal more flavor than their meat might indicate and may also be used as a partial basis for brown stock. In fact all of these cuts can be used by themselves or in combinations.

Shin

Shin meat contains a great deal of elastic connective tissue. The flavor is rich. The shin or knuckle bones from the hindquarter contain more marrow than those from the forequarter, but either make ideal additions to the stockpot. The marrow makes good eating too.

Short ribs and their meat make a stock that is not quite as dark and rich as that of shin meat, but still delicious and full of body.

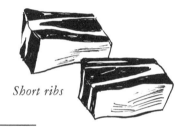

Short ribs

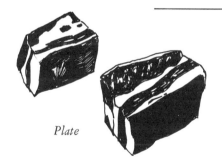

Plate

Plate beef meat is rich, but much fatter than that of short ribs. It is also much less expensive. In fact, plate meat costs the least of all the cuts listed for stock, and the fat it renders into the broth can be skimmed off easily after it has cooled.

Chuck neck bones contain a little meat and are the very best for bouillon. They lend a hearty flavor to a stock and have the virtue of very little fat.

Beef brisket bones that contain a little meat will yield a sweet delicate stock. They are very inexpensive. You can buy a whole pile for an investment of less than 50 cents.

For white stock use the meat and bones from the breast, shank or shin, heel, and shoulder of veal. Use any veal leftovers too. The remains of your last veal dinner may, in fact, be just as good.

The options are always open with a stockpot. The cuts listed above are but a few limited examples. Once you get the knack of making a stock, you may even find yourself planning your menu around the bonus of a bone or two. Consider the following list if you do:

	For your dinner	*For your stockpot*
BEEF	chuck steak	blade, chine bones
	plate pot roast	brisket bones, cartilage
	short ribs	rib bones
	sirloin steak	wedge bones
	cross rib roast	finger, knuckle bones, cartilage
	Porterhouse steak	t-bone with chine
	rib roast	feather bones
	rump roast	aitch, knuckle bones
VEAL	rump roast	aitch, knuckle bones
	loin roast	back bones
	shoulder roast	blade, arm bones
	boned breast	breast bones, short ribs
	stew	shank bones
	shoulder chop	blade bones

. . . and for those uniquely flavored stocks . . .

LAMB	leg roast	knuckle, aitch bones
	shoulder roast	blade, neck, chine bones
	loin chop	back bone
	rib chop	feather bones
	loin roast	back bones
	shank of lamb	shank bones
PORK AND HAM	shoulder	knuckle bone
	Boston butt	blade bone
	fresh ham	leg, aitch bones
	shank	knuckle, leg bones
	butt end of ham	aitch, spool bones
	pork loin	aitch, blade, back, rib bones

Part III

Pork
and Ham

Chapter 9

◦┼◦┼◦┼

The Whole Hog

standards of quality...the primal cuts

Today when a pig goes to market, he can go any one of a number of ways. About 33 per cent of the pork produced in the United States is marketed fresh, and the remainder is sold smoked, pickled, cured, rendered into lard, or processed under a brand-name label. Pork is available the year round, but the season of greatest abundance is the late fall and early winter months.

Fresh pork is the richest known food source of thiamine (Vitamin B_1). Its lean is no higher in fat than the lean of beef, and in comparison it costs very little. Yet today in America we eat only half the amount of pork that we eat of beef, which is about the same amount we ate at the turn of the century. The methods used for raising pork have changed a good deal since then, however. The sty has been replaced by scientific pig farms where hogs are fed high protein diets.

The U.S. Department of Agriculture grades of pork are U.S. No. 1, U.S. No. 2, and U.S. No. 3. These grades cannot be compared to those of beef and lamb and veal. There are fewer grades and less discrepancy between them. The amount of fat is actually the determining factor of what grade a porker will rate. Since nearly all the hogs prepared for market are five or six months old, weigh between 200 and 220 pounds, and have been well fed—excessive or deficient fat may be the only difference between grade 1 and grade 3. The expensive exception is baby suckling pig which weighs in at 20 to 35 pounds, and is sold intact except for its innards.

When purchasing pork, the thing to look for is lean, young cuts. They can be recognized by their color. Grayish-pink indicates a cut that is young, as does a pure white covering of fat. Both the lean and fat are firm. The older cuts have a deeper red coloration and are often flabby, soft, and unshapely. Another way to determine the youthfulness of pork is to examine the bones. In young pork the bones are soft with a slight tinge of red. In older cuts they are brittle and white.

One of the rules of fresh pork is that it must always be cooked well done. The reason for this is simple. Hog muscles are sometimes infested with the parasite trichina which causes trichinosis in humans. Thorough cooking destroys the parasite and totally eliminates the possiblity of infection. In years past it was believed that in order to destroy trichinae, pork should be cooked until the internal temperature reached 185°F. Newer cooking studies prove, however, that an internal temperature of 170° is sufficient to kill the parasite and produces a juicier cooked pork product with less shrinkage.

PORK PRIMAL CUTS

Fresh pork is generally received by the processor or retailer already divided into primal cuts. There are seven primal cuts of pork.

1. Leg of pork
2. Loin of pork
3. Pork belly
4. Breast of pork
5. Pork shoulder
6. Pork jowl
7. Pork feet

1. **Leg of pork,** which is often referred to as fresh ham, represents 18 per cent of the hog and yields the greatest amount of solid, lean meat for roasts and steaks. The best leg of pork or fresh ham has skin that is smooth, white, and free of abrasives. When cooked it has a very delicate flavor. It may be purchased whole or de-boned for stuffing. More

Leg of pork

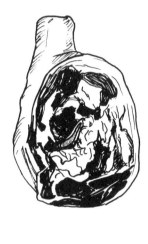

often, however, a 12- to 16-pound leg is divided into two roasts: a 6- to 8-pound fresh ham butt end, and a 6- to 8-pound fresh ham shank end. Some meat cutters even stretch these two roasts further—particularly during sales at the supermarket—by cutting a slice or two out of the center and selling them as fresh ham steaks. This is the best time to purchase fresh ham steaks, but when purchasing a half leg of pork or fresh ham it is best to specify full butt or full shank or literally you may be short-shanked.

In the supermarket, you can distinguish a full butt from one with a center slice removed by careful examination. In a full butt, the upper bone (or aitchbone) will be fully recessed into the flesh and only the small spool bone will be visible in the center. In the case of the shank half, the center spool bone will appear elongated and flat if the center slices have been tampered with.

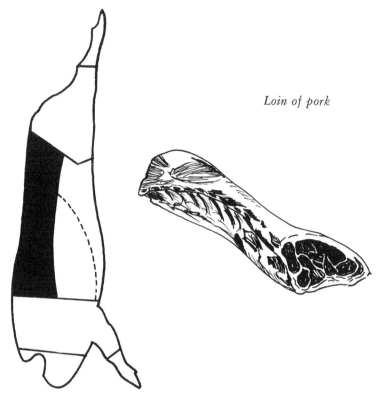

Loin of pork

2. **Loin of pork** consists of the greater piece of the backbone and encompasses part of the shoulder. Young whole pork loin weighs from 10 to 14 pounds and represents about 15 per cent of the animal. It may be purchased whole or divided into roasts and chops of varying quality. The cuts from the extreme loin end contain a great deal of bone. Those from the extreme rib end have more fat. Both cost less than the center cuts, but they yield less tender, tasty meat. The center cut, which is about 8 pounds of a 14-pound loin, corresponds roughly to the rib section of the beef. It yields roasts and chops that are far meatier for your money.

Whatever section of the loin you decide to purchase, it will be more delicious if it is cut from a small, young hog. A palate-pleasing rule to follow is to look for the smaller chop. It is bound to come from a younger animal. In contrast, large chops usually come from a more mature animal.

Pork belly

3. **Pork belly** represents almost 18 per cent of the animal and adapts well to smoking and curing. For that reason, meat processors snap up most of the fresh pork belly and turn it into salt pork and bacon. The best bellies are those that yield the top-grade bacon. They are a uniform 1 to 1½ inches wide and have pronounced streaks of lean intermingled within the fat. Fresh pork belly can be purchased in some stores and specialty shops, and sometimes serves as the basis of home-made sausage.

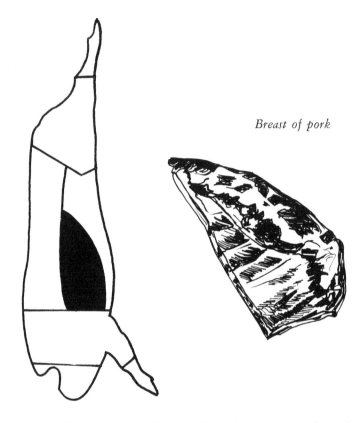

Breast of pork

4. **Breast of pork** is the primal cut that gives us spare ribs. It is only 3 per cent of the entire animal. The butcher usually cuts and trims the ribs for the consumer from a whole rib section that weighs anywhere from 2 to 5 pounds and contains up to thirteen ribs. These whole rib sections are called "slabs" of spareribs, and their weight is important. The ribs cut from 2-pound slabs are the most choice, since they come from a younger animal and have less bone and more tender meat. They are, however, in short supply and quite expensive. The best bone-to-meat ratio for the money is offered by ribs cut from 3-pound slabs. The meat is lean and tender, and the bones acceptably small. The heaviest, most mature ribs can be recognized by their well-developed, cartilaginous bone end. Unless these older, bony, spareribs are bought on special at a reduced cost, they are no bargain. The consumer is paying for the inordinate amount of bone.

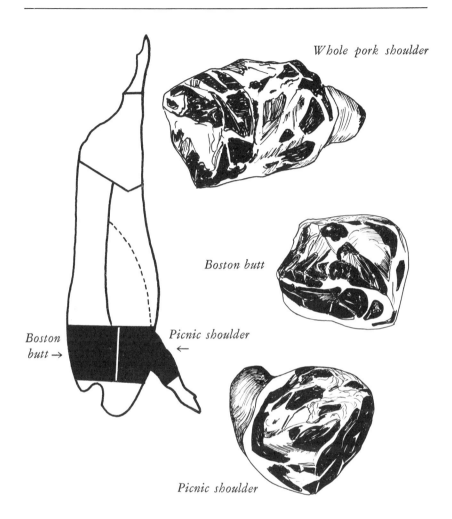

Whole pork shoulder

Boston butt

Boston butt →

Picnic shoulder
←

Picnic shoulder

5. **The pork shoulder** weighs from 12 to 16 pounds and represents 15 per cent of the animal. Visually, the shoulder resembles the fresh ham of the leg of pork, but the meat it yields is not as tender and lean. The shoulder is sometimes sold whole, in which case it is better if de-boned because the complicated interior bone structure makes for tricky carving. The shoulder can also be divided into fresh (or Boston) butt, picnic (or Calia), hocks, and arm steaks.

Pork jowl

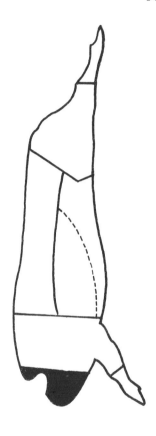

Jowl bacon

6. **Pork jowl** by and large is used by processors, mainly to make jowl bacon. It is also called for in certain ethnic dishes.

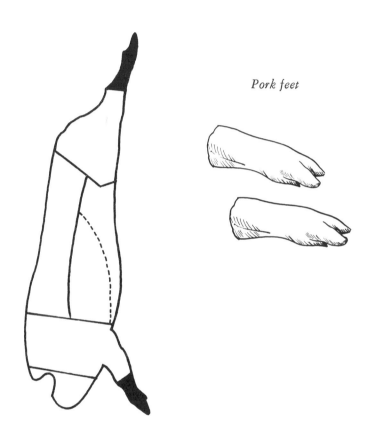

Pork feet

7. **Fresh pork feet** may be purchased whole or subdivided into knuckles and hocks. This division should not be attempted with a cleaver, however, as the brittle bone will splinter into the meat. Many cooks use pork feet for flavoring recipes. Many of the feet are also purchased by processors for pickling.

Chapter 10

❂❂❂❂❂

Pork—The Retail Cuts

roasts... chops... steaks... barbecue

Y̲ou can purchase almost any part of a pig from his snout to his hind feet knuckles. The fat back, located above the loin area, for instance, is combined with belly trimmings and rendered into that delicately flavored, inexpensive shortening: lard. Fat back is also used by butchers for wrapping (or barding) lean beef and veal roasts, for larding meats, and for flavoring. Barding and larding are methods of adding fat to meat to make it more moist and tender. Some pork and bean manufacturers also put a bit of fat back into every can. The homemaker can purchase it either cut from fresh slabs, or vacuum packed in round or cylindrical packages referred to as chubs of ¾ to 1½ pounds.

In addition, the intestines are washed and become chitterlings. The skin, snout, ears, and tail are also used in certain styles of ethnic cooking. The cuts discussed here are the fresh pork roasts, chops, steaks, and budget barbecue cuts.

SEVEN BASIC PORK ROASTS

A pork roast can be the basis of an inexpensive and delicious meal. And it is even easier to cook to perfection than a beef roast because it is invariably cooked well done. And, the pork's smooth white surface fat melts and provides sufficient moisture to keep it from drying out while it's roasting, so it doesn't require basting or fuss.

Which pork roast should you buy? There are many from which to choose. Seven are listed below in descending order of tenderness, leanness, and the uniform shape of the cut:

1. Center-cut loin
2. Center-cut rib
3. Full loin half
4. Full rib half
5. Fresh ham (or leg of pork)
6. Fresh shoulder of pork
7. Fresh pork butt

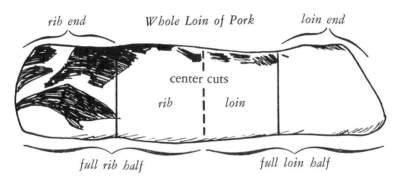

A whole 10- to 14-pound loin of pork
(center cuts = 8 pounds)

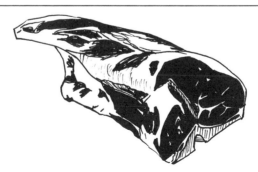

1. **Center-cut loin** is the ultimate pork roast. It weighs about 4 pounds, and as it contains the tenderloin (the same as the beef tender-

loin) it is meaty from end to end. This cut does have a somewhat complicated bone 'structure, however. Therefore, it is best to have the small back bone either cracked or completely removed to make carving easier. This roast can serve from four to five people.

Center-cut rib roast

2. **Center-cut rib** also weighs about 4 pounds. The meat is as tender as that of the center loin, but because the center rib does not contain the tenderloin there is just not as much of it, and it will take nearly a pound of meat to provide one serving. The rib is a little fattier than the center-cut loin and can be carved easily into uniform chops after roasting. A trimmed and shaped center-cut rib is a crown of pork roast.

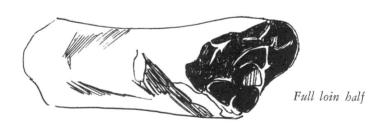

Full loin half

3. **The full loin half** weighs from 2 to 3 pounds more than the center-cut loin of pork but is generally sold for less per pound. It includes the less desirable bony loin end close to the hip. But it is an exceptionally lean and tender cut which can be a superb roast, especially if it is completely de-boned. The full loin half roast will provide six to eight servings.

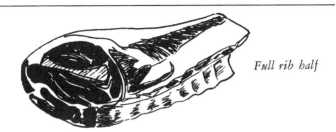

Full rib half

4. **The full rib half** weighs from 5 to 7 pounds and will feed as many people. It is a good choice for those who favor the flavor of sweet bones. Like the full loin half, the full rib half encompasses the less desirable and fattier rib end nearer the shoulder of pork, so therefore sells for less a pound than the smaller, more select center-cut rib roast. This roast is at its best when the bottom of the chine bone has been removed to make for easy carving.

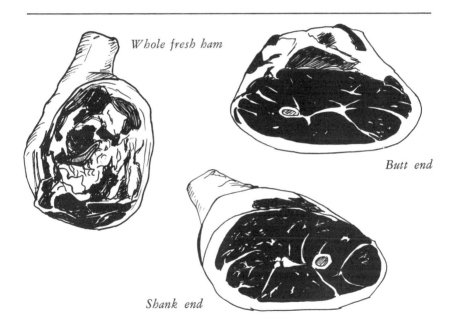

Whole fresh ham

Butt end

Shank end

5. **Fresh ham (or leg of pork)** has firm, lean meat and weighs from 12 to 16 pounds whole. Its quality is evidenced by a grayish pink color, broad, deep shape, creamy white surface fat, and well-marbled face. This party-sized roast will feed eighteen to twenty people or more and is also sold de-boned. The fresh ham can, however, be purchased split

into two smaller roasts—a full shank end and a full butt end. (See Chapter 9, "The Whole Hog.") Of the two, the fresh ham butt end is probably the better value because it is leaner and meatier although harder to carve than the shank end. But that problem can be solved if the small hip bone (or aitchbone) is removed by the meat cutter. The full shank end of a fresh ham is less meaty but easier to carve. Of the two sections, the shank end is most likely to be found at supermarket sales. You might want to consider buying a whole fresh ham and having it divided into several meals. Save the butt end for roasting and have the shank end sliced into steaks.

Picnic shoulder

6. **Fresh shoulder of pork,** often referred to as fresh picnic or calia ham, weighs from 5 to 8 pounds and although not as lean as the loin or fresh ham, is, nevertheless, a very flavorful, inexpensive fresh pork roast. The shoulder of pork can be roasted as is, but the complicated bone structure makes carving difficult. If trimmed of surface fat, de-boned, rolled, and tied, the shoulder roast is most attractive and easy to carve. Either way the shoulder of pork is a fine value for shoppers. A pound of boneless shoulder roast will feed two to three people, but with the bone in you will need nearly a pound per serving.

7. **Boston butt** is also called shoulder butt and is, in fact, part of the shoulder. Most often this cut is cured and smoked, but fresh butt is among the most distinctively flavored pork roasts. However, it is extremely fatty and contains the most waste of all the roasts listed.

Firm, small, fresh pork butts, weighing about 4 to 6 pounds, with

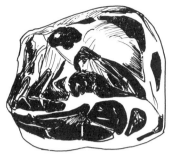

Boston butt

clear white exterior fat are the best quality to buy. As a rule, these will have less fat content than the larger butts, and therefore less waste and more value. Because they contain a good deal of bone, it will take ¾ to 1 pound to provide a serving. For easiest carving, purchase this cut with the blade bone removed.

PORK CHOPS AND STEAKS

Like most meats, the quality of a pork chop or steak is determined by the area from which it is cut. Since there are only four different kinds of pork chops cut from a hog, selecting a quality chop—or a bargain chop—should be a fairly simple chore. This is not always the case, however. If you go to a butcher, you can name your cut and expect to get it, but in the supermarket, uniform quality pork chops can be a little harder to come by.

A recent merchandising gimmick, for instance, is the "quartered pork loin," presliced into 9 to 12 chops and prewrapped for the consumer. At first glimpse this fairly low-priced package may seem like a bargain. But don't be misled. It is not. A "quarter loin" could be from any section of the whole pork loin, and you can be certain that the largest majority of the presliced chops will be from the less desirable end cuts. The package will probably contain a few center-cut chops, but not enough to justify the overall cost of the entire quarter loin. If you want quality you should specify center-cut loin or rib pork chops. But if you are in the mood for a bargain, it is far better to look for and specify end cuts. The saving will be about 20 per cent over the cost of the presliced, prepackaged quarter loin, and you will virtually be getting the same pork chops that the package offered.

A pork steak can also be a bargain—one that few consumers take advantage of. Yet a pork steak is a delicious and attractive addition to any meat repertoire. It can be fried, braised or barbecued, and is an inexpensive, extra-meal bonus cut from a fresh ham butt or shank end pork roast.

Listed below are four pork chops and three pork steaks in descending order of overall quality:

1. Center-cut loin chop
2. Center-cut rib chop
3. Loin end chop
4. Rib end chop
5. Fresh ham steak
6. Shoulder arm steak
7. Blade pork steak

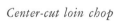

Center-cut loin chop

1. **A center-cut loin chop,** which encompasses part of the tenderloin, is the finest you can buy, and it is the most expensive. The cost is consistently high because there are only 10 to 12 center loin chops in an entire hog. The demand is always great for perfectly tender, uniform chops. When you consider that the center loin chop has more lean meat and less bone and waste, however, the price factor does compare favorably to that of the lower-cost cuts from the loin ends.

Pocketing a loin chop

Buy one loin chop per serving. If you wish to stuff the center-cut loin chop, select those sliced one to two inches thick. Place the chop flat, and carefully insert a small, sharp knife about a quarter of an inch into the flesh at the tip of the chop. Then slice a small flap from the tip to the base.

2. **The center-cut rib chop** differs very little in quality or flavor from the center-cut loin chop. It does not, however, contain that delicious morsel of tenderloin. The additional rib bone is really an asset to the chop because it helps conduct heat and

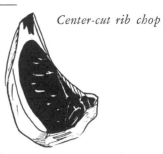

Center-cut rib chop

expedites cooking. Center-cut rib chops from a lean, young hog are the most desirable ones to buy. Depending upon thickness, select one to two center-cut rib chops per serving.

Center-cut rib chops lend themselves nicely to "Frenching." To make French pork chops, the meat is shaved away from the lower end of the chop to completely expose the rib bone. After the chop is cooked a fancy paper frill is sometimes added to garnish the bare tip of the bone.

Loin end chop

3. **The loin end chop** is often referred to as a sirloin or hip pork chop. It is less desirable than chops cut from the center loin because it contains a great deal of bone and less lean meat. And because of this extra bone, the end-cut loin chop is priced about 20 to 30 cents less per pound than the center cuts. Therefore it is an especially good buy if purchased at a supermarket special. Depending upon thickness you will need one to two end-cut loin chops per serving.

4. **The rib end chop,** or shoulder pork chop, is very tender but contains a great deal of fat so that it is always priced lower. Look for slender rib end chops. The wider chops come from older hogs and will have the most fat and harder, broader bones. Depending upon thickness one or two chops will be required per serving.

Rib end chop

Fresh ham steak

5. **The fresh ham steak** is lean and tender and those cut from the shank end of the whole fresh ham will be particularly uniform in size and attractive. The best time to purchase fresh ham steaks is when fresh ham is on sale at the supermarket. Select a steak about ¾ to 1 inch thick. One steak will serve two generously—or even three people.

Arm steak

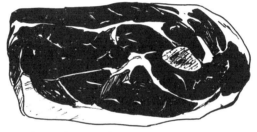

6. **The arm or shoulder steak** of pork is not as lean and tender as the fresh ham steak, but it is generally far less expensive. The way to really take advantage of this low-cost cut is to purchase an entire 5- to 8-pound shoulder. The butcher can cut it into as many as ten separate steaks providing as many servings—a very tasty way to save money.

Blade steak

7. **The blade pork steak** could be the basis for a delicious and unusual budget barbecue. The fresh shoulder does contain a good deal of gristle, fat, and a wide section of blade bone, but the cost is very nominal. An entire 3½- to 5-pound butt can yield up to ten to twelve economy-sized steak servings that are even less expensive than hamburgers.

PORK TO BARBECUE

If you like pork and are on a tight budget, you can barbecue—so many of the cuts cost so little. The mild, delicate flavor of pork lends itself so deliciously to barbecue saucery, that even the cheapest cuts prepared barbecue style can be a delectable treat for a big gathering.

Everyone knows the tasty goodness of spareribs hot from the grill. Listed below are six other inexpensive barbecue ideas that capitalize on pork's versatility:

1. Country style spareribs
2. Pork butt steak
3. Pork cutlets
4. Butt end fresh ham steak
5. End-cut pork chops
6. Shoulder or arm steak

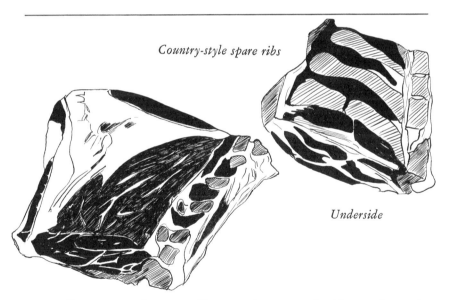

Country-style spare ribs

Underside

1. **Country style spareribs** are the meaty backbones from the shoulder end of the pork loin. They cost up to 20 per cent less per pound than spareribs cut from the breast of pork, yet they contain more meat and are very delicious. One pound yields two to three rib cuts—enough for about one or one and a half servings.

Pork butt steak

2. **Pork butt steak,** cut from the Boston butt as it is sometimes called, is a tender and flavorful cut that is ideal for barbecuing because of its natural marbling. It may be prepared and eaten with the bone in, or de-boned and eaten on a bun. These little boneless cuts of pork steak are sometimes known as pork "tenderettes." A whole Boston butt will provide about ten of these boneless pork steaks. Figure one to two per serving depending upon appetites.

3. **Pork cutlets** from the shank end of the fresh ham are extremely inexpensive. This cut is particularly adapted to the intense heat of barbecue cookery because it is well-marbled and will not dry out. A whole shank end of fresh ham will make enough small cutlets to feed from six and up to twelve depending upon how thickly they are cut and how they are served.

4. **Butt end fresh ham steak** is both thrifty and meaty. A small bone in the steak holds the meat together firmly, and it has very little waste. A steak ¾ to 1 inch thick will serve two to three generously.

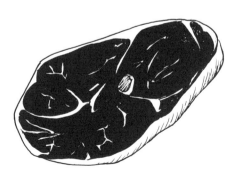

Butt end fresh ham steak

5. **End-cut pork chops** from either end of the loin are inexpensive and take well to barbecue cookery. Depending on appetites, figure one to two chops per serving.

End-cut pork chops

6. **Shoulder or arm shoulder steak** is a natural for barbecuing. A single, 5- to 8-pound whole shoulder will provide up to ten of these steaks. When the individual steaks are barbecued, a small round bone pops out to tell you they are ready to eat. A very tender, tasty, most economical treat. One steak will more than suffice for a serving.

Arm steak

Chapter 11

❖❖❖❖

Ham and Smoked Pork

the curing process... convenience, taste, size...
selecting quality bacon

A piece of pork becomes a piece of ham after it has been thoroughly processed. Originally, curing, smoking and aging pork were simply to preserve the meat. Years ago, prior to modern refrigeration, many a pig would have passed the point of palatability if it were not for "the cure"—a process which largely involved salt, and produced ham, salt pork, and bacon.

According to the *Larousse Gastronomique,* the curing of pork can be traced to the ancient Gauls. They not only knew how to raise strong, healthy pigs in profusion but also loved to eat their meat and thus became masters of the art of pork preservation. They became so skillful that soon their reputation spread, and before long they were the major ham distributors to all of ancient Rome.

Taste, rather than preservation, is of primary concern today, and taste governs the methods used in producing ham, smoked pork, and bacon. Many a manufacturer keeps his special recipe for processing ham hush-hush—strictly a well-kept secret.

Although the grading and primal cutting of ham is virtually the same as it is for pork, the method of raising pigs for their final product and subtle flavor variation may differ from area to area. For example, some hogs eat only acorns, others only peaches. Smithfield hams have long been produced from peanut-fed hogs raised in the peanut belt of Virginia and North Carolina. At the turn of the century these hams were so widely imitated that by the 1920's the State of Virginia was forced to pass a statute stating that the only real Smithfields were those produced from peanut-fed hogs processed in the town of Smithfield, Virginia. The greatest majority of hogs, however—nearly four-fifths— are corn-fed.

Over 60 per cent of the pork produced today undergoes some sort of processing. Although the methods have been updated, making pork into ham and bacon still takes the same essential steps that it did years ago: curing, and sometimes smoking, and aging.

Curing is a process which utilizes salt to retard the growth of bacteria in pork. To offset the hardening effect of the salt, sugar is used too. "The cure" itself may take one or a combination of three forms. It can be "dry"—the method whereby salt and a mixture of spices are rubbed directly into the surface of the meat. Secondly, the pork may be totally immersed and kept in a brine solution to which garlic, pepper, and spices may have been added. And finally, by far the most sophisticated curing method: a brine solution is injected into the pork itself. Whichever method is used, the length of time required to cure a ham depends largely upon the amount of time it takes the spices or liquid solution to diffuse sufficiently throughout the meat. This, of course, depends upon the thickness of the cut and also upon the temperature at which it was stored. In any case, the brine content of any ham cannot exceed 13 per cent according to U.S. Department of Agriculture standards.

Smoking, the second step employed in making ham and bacon, is achieved in a hot, airtight smokehouse. The mildest flavored, precooked canned hams eliminate the smoking step altogether. Smoking is the process that adds flavor and color—gives real "personality" to the meat. How flavorful and colorful a ham will be is dependent on the sort of wood used to produce the smoke—hickory, beechwood, juniper twigs, berries, or plain sawdust—and how long the meat is smoked. The United States Department of Agriculture requires that any ham labeled "ready-to-eat" must be kept at an internal temperature of 140° F. for not less than thirty minutes. That amount of time would produce a very juicy, lightly smoked, mildly flavored product. Heavy, long smoking produces a drier product and the richer flavor and color characteristic of those specialty ham products: Prosciutto, Westphalian, and Smithfield. (The latter are smoked for about thirty days.)

After the smoking process, the ham may or may not be set aside and hung to age for a period of up to a year. The more expensive hams will usually be the ones that have undergone complicated processing, a high degree of smoking, and longer aging. These steps not only tend to cause the ham to shrink, dehydrate, and lose weight—they also cause the price to escalate.

In the tropical countries where there is no refrigeration, it is necessary that hams be given what is called a "tropic cure." The hams are dehydrated—dried out completely. Slices from a tropic-cured ham resemble chipped beef.

Traditionally, the only meat referred to as a "ham" was that pork cured from the leg of a hog. Today, however, a variety of cuts are known as ham. One can purchase, for instance, "picnic hams" cut from the shoulder of the hog or a "daisy ham" from the cured pork butt. Most of these pork products have a taste—and sometimes even a look—similar to that of traditional ham, but they are not as lean nor are they as attractive. But neither are they as expensive. The leanest, meatiest, most elegant and impressive of all the ham "cuts" is still the traditional 10- to 14-pound cured, smoked, aged leg of pork with the bone intact.

Selecting a quality ham is somewhat more difficult than choosing other cuts of meat. Many of the hams available are canned, vacuum packed, or bagged, making it impossible for a customer to see the product. He must, to a large extent, make his selection on the basis of the printed material on the label and those priorities of convenience, taste, and size.

CONVENIENCE

Would you like a tasty, easy-to-fix ham meal or an elegant one that could require a number of days to prepare? Today the option is yours. You may purchase a ham labeled either one of two ways: "ready-to-eat" or "cook before eating."

If a ham is labeled "ready-to-eat" (or "fully cooked," "ready to serve," "tender 'n ready"), you may do just that: eat it cold, just as it is, without further cooking. It will, however, taste and look better if some condiments and spices are added to the outside surface and the ham is reheated and glazed before it is served. The labels on most ready-to-eat hams will instruct you how to further prepare them quite easily.

The old-fashioned "cook before eating" or the "country style" hams are far less common and usually sold whole. These hams require a far more complicated cooking procedure. Some of them must be soaked for twelve hours or more (or par blanched), then scrubbed, simmered, skinned, spiced, reheated, and glazed if one so desires before they are set on the table. This kind of ham will be perfectly delicious, smokier in flavor and aroma, and less expensive per pound than the "fully cooked" type of ham. But, obviously, a potential customer will want to consider the extra labor involved on his part before deciding which of these two types to buy. Generally the hams found in cans will be of the "ready to eat" type. Follow your favorite recipe, whichever type you choose.

Whether a ham is skinless and boneless is yet another aspect to consider. As with other bone-in meat roasts, the bone will act as a conductor of heat. Those ham roasts containing a bone will, therefore, cook more rapidly per pound than those without bones, and will also generally be juicier and tastier. The boneless hams, although they may require five more minutes per pound to cook, will be easier to carve. And those hams without the hard outer rind removed will also require skinning prior to being glazed and served.

Canned hams also offer another kind of instant convenience—the kind one needs when unexpected guests arrive. Some of the larger canned hams can be stored successfully in the refrigerator for up to three months or more. The Polish and Dutch—but particularly the Danish canned hams—are of superior quality to our own domestic canned hams. They cost more, about 20 cents more per pound. Some of the smaller canned hams, including many of the imported hams under 3 pounds, can be stored without refrigeration. During their canning process, these hams have actually been sterilized—in essence really overcooked so that the texture has broken down. They are not the same quality as the larger hams, therefore, nor are they the same cut. Check the label to make sure whether or not a canned ham requires refrigeration.

TASTE

How do you like your ham? Smoky? Mild? Salty? Preglazed? The flavor of ham most preferable to you is, of course, strictly a matter of taste. Those hams mentioned earlier—Smithfield, Westphalian, Prosciutto —have very distinctive flavors considered by some to be a great delicacy, while others find them far too salty or smoky to eat. If you have never sampled one of these specialty hams before, it is suggested that you purchase a quarter pound or less, sliced at a specialty shop or delicatessen, and taste it before venturing into a big investment. These hams are generally sliced paper-thin so that an even thriftier way to order them would be by the slice, according to the amount you estimate you will need. In fact, this rule can apply to all cold cuts. And, also, don't hesitate to ask for a sample taste before buying.

Most hams available today are sold under a manufacturer's label. If you have not yet discovered your favorite brand of ham, the best way to do so is by trial and error. Simply taste test a few until you find the one that best suits you. Then you can be certain that you and your taste can rely on that particular brand label. A good point to keep in mind is that a whole ham will generally be more flavorful and juicy than anything smaller cut from it. The whole cut will simply retain more of its natural juices.

Canned hams will often have a milder flavor because many of them have not been smoked, but merely steam cooked or sterilized in huge factory kitchens. There are a few exceptions, however, so be sure to look at the label. It generally indicates whether or not the flavor of the ham is smoky.

SIZE

How much will you need? How many are you serving? Do you want leftovers or not? (Unsliced, leftover ham, well covered in the refrigerator, will last for a week or more and definitely comes in handy for sandwiches, soups, casseroles, and omelets.)

When it comes to selecting the size of a ham, the options are open. Ham is cut in every size and shape imaginable. You may choose anything from a small, canned chunk to a large elegant roast and even individual or small family ham steaks are available. The amount of meat you need and how fancy you want it to be will probably dictate which of the seven basic types of ham listed below you will ultimately select. Listed in descending order of quality and quantity of solid lean meat a cut yields for your money, they are:

1. Whole ham (with bone)
2. Whole butt end of ham
3. Whole shank end of ham
4. Whole boneless ham
5. Smoked half loin of pork
6. Smoked shoulder of pork
7. Smoked butt

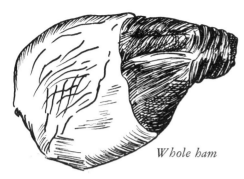

Whole ham

1. **A whole ham** (with the bone) is the least wasteful, most distinctively flavored cut you can choose. In order to get the best quality, select a plump ham with a short, stubby shank end. According to brand and size, a whole 10- to 14-pound ham can feed twelve to eighteen people generously, and you can probably count on some leftovers too.

2. **The butt end of ham** is a convenient size for average families. It is harder to carve than the shank half, but usually meatier. When purchasing, make sure the label specifies "whole" butt half, or some of the center slices may have been "stolen" for ham steaks. For variety, you may want to have those center slices removed as individual ham steaks, and use the rest for baking.

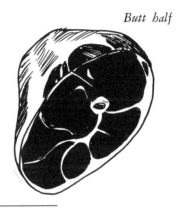

Butt half

Shank half

3. **The shank end of ham** has meat as tasty as the butt half, and it is easier to carve. Because it has less meat it is usually less expensive. Again, be certain that the label specifies "whole" shank half so that you are not missing some of the center slices. A 5- to 7-pound shank half of ham will feed six to eight people. You may want to have the extreme shank end removed for making and flavoring soups and vegetables.

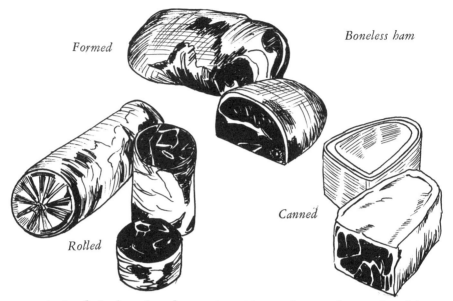

Formed

Boneless ham

Canned

Rolled

4. **A whole boneless ham**—8 to 12 pounds—can be purchased in almost any shape. After the bone has been removed the meat is custom shaped and packaged in cans or airtight wrappers. The boneless ham will lack the flavor of a ham with the bone intact, and cost 20 to 30 per cent more per pound. The entire ham will provide enough meat to feed a few dozen people—or more—but you may also purchase boneless ham in halves or chunks of varying sizes. Allow about ¼ to ⅓ pound per serving.

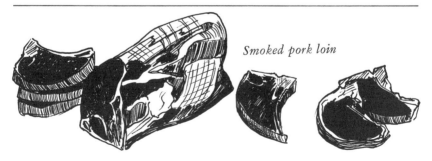

Smoked pork loin

5. **A smoked half loin of pork** with a bone will weigh 5 to 6 pounds. Although it is a fairly unusual cut and somewhat expensive, it has gained some popularity lately because of its delicious flavor. The meat is as tender as that of regular ham, but it is drier. Because of the narrow circumference of the cut, the loin tends to dry out faster during

the smoking process. A half loin with the bone should feed 6 to 8 people. This cut will be practically impossible to carve unless the chine bone has been cracked between every rib. After the chine is cracked, the roast smoked loin will easily carve into neat little chops when you are serving. Often the loin is boned out and cured by a slightly different method in order to make Canadian bacon.

Smoked shoulder

6. **A smoked shoulder of pork,** also known as a "picnic ham" or "Calais ham," is priced much lower than the other choices. This is a somewhat wasteful cut that contains more fat, bone, and skin in proportion to its lean meat. You can figure that it will take almost a pound of smoked pork shoulder to feed each person when purchasing a roast. The whole shoulder will weigh 5 to 8 pounds, but perhaps the best way to purchase this difficult-to-carve cut is to have it machine sliced at the store into individual steaks, and then put it back together, tie it, and bake it "ready sliced."

Smoked butt

7. **The smoked butt** is a boneless cut from the neck and the shoulder of the hog. It is also known by various other names: Boston butt, smoked tenderloin, and daisy ham. The butt meat is tender and delicious, but can be extremely fatty. A long, narrow selection of about 2 to 2½ pounds will probably be leaner than a short plump one and will feed three to four people. The smoked butt also offers a delicious and economical alternative to bacon.

BACON

Bacon, although similar to smoked pork and ham, has other qualities for the consumer to consider. Smoked bacon is generally cut from the belly of the hog, and is by definition a very fatty cut—one that is extravagant and extremely wasteful. The leaner the bacon, the better its quality—but almost 80 per cent of each strip of bacon is comprised of fat. Bacon does have a unique, delicious flavor and aroma, however, and, when cooked, such a crispy texture that it has always been thought of as the classic breakfast treat to go along with eggs.

The best bacon you can bring home is firm to the touch—never flaccid. It has a streak or two of bright pink lean meat within the fat and is uniformly thick from end to end. If there is moisture inside the vacuum pack, or the bacon has a white, slimy appearance, it has probably been around too long and will tend to be chewy rather than crisp and hardly as delicious as fresh bacon when it is cooked.

Bacon does, in fact, lose its distinct aroma and flavor quite rapidly after it has been opened. It is recommended, therefore, that no more than a week's supply be purchased at any one time. Although bacon may be frozen for short periods of time when it is absolutely necessary, some of its flavor will be lost, and the spatter caused during cooking by the thawing ice crystals or accumulated moisture make freezing extremely undesirable.

The imported bacons on the market from Canada, Ireland, and Denmark are usually not the product of corn-fed hogs as most of our bacon is in America. Rather, it is cut from hogs who have eaten alfalfa, oats, or bran. Since corn produces fat, the imported brands tend to be less fatty and vary slightly in taste due to the different cures. Canadian bacon is unique. It is cut from the loin of pork rather than the belly and is meatier than American bacon.

Most bacon may be purchased in one of five styles:

1. Slab
2. Regular sliced
3. Thin sliced
4. Thick sliced
5. Ends and pieces

---

1. **Unsliced slab bacon** is not as commonly found in most markets as the sliced variety because it is not as convenient. It must be sliced by

Slab bacon

hand before it is cooked. And it has a thick outer rind. This rind can be removed before the bacon is cooked, or afterwards—European style— to help prevent the bacon from curling. Slab bacon is less expensive than sliced bacon, and will last much longer when stored unsliced under refrigeration. The most important asset to look for in slab bacon is equal thickness from end to end. The top fatty surface should have observable pink streaks of lean, and the fat should be pure white.

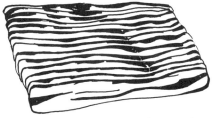

Sliced bacon

2. **Regular-sliced bacon** is a good buy and more attractive than slab bacon because its slices will be uniformly cut. A pound will contain anywhere from eighteen to twenty-two slices.

3. **Thin-sliced bacon** fries more quickly than regular sliced, and it offers more—but smaller—portions to the pound. Each pound of thin-sliced bacon averages thirty-five to forty slices.

4. **Thick-sliced bacon** is sometimes called "country style" and is delicious but not as crisp as regular and thin sliced. Each pound will contain twelve to sixteen slices which will take a little longer to cook than the other types.

5. **Ends and pieces** are sometimes featured for sale at a bargain price. Although they are hardly as attractive as the other styles of bacon, they are very useful as garnishes for salads, casseroles, and other dishes.

A word of caution for those who insist on the best quality of bacon: if you are accustomed to purchasing the film-wrapped, vacuum-packed type of bacon, as most people are, one sure way to determine if you are getting the top-graded variety of bacon is by perusing the packer's label. Most packers provide three or four different quality selections of bacon. If the packer's name is featured prominently on the label in bold letters or is part of the name of the bacon product you can be fairly certain that the contents of the package are the top of the line—the best that packer has to offer. The meat will be cut from the better young hogs, and the slices will meet rigid standards of uniformity and attractiveness and will most likely meet with your approval. It's a matter of pride to the packer. To feature his name on the label is his own peculiar stamp of approval.

If, on the other hand, the bacon is merely the second or third best a packer has to offer, it will probably be priced considerably lower. His name will usually appear on an obscure part of the label in tiny, fine print, and the label will merely indicate, as guardedly as the law allows, that this bacon is a product of that particular packing company. The bacon slices within the wrapper may be irregularly shaped and not as attractive as those of the packer's best. The meat may also be of a somewhat coarser texture, but the bacon will be just as tasty, however.

By the time this reaches print, bacon manufacturers will probably be complying with proposed new regulations designed to protect the consumer. Under these regulations, 70 per cent of the representative bacon product must be visible through the packaging—rather than the tiny portion visible in current packaging. The prospective customer will benefit by being able to more accurately determine the actual ratio of fat to lean.

To separate the slices most easily, simply flex the package of bacon while the slices are intact, or remove the bacon from the refrigerator five or ten minutes prior to cooking it. Then it may be prepared in any one of three ways: frying, baking, or broiling.

To fry, place the slices in a cold pan, and turn them often to achieve uniform crispness. It is not necessary to pour off the fat while cooking unless the spatter gets too much for you.

To bake, preheat the oven to 400° F. and place the bacon slices on a wire rack in a shallow pan to catch the drippings. Baking will take ten to fifteen minutes.

If you wish to broil bacon, be sure to keep the slices three inches from the source of the heat and turn the slices at least once. Always drain bacon slices on absorbent paper before serving—or better yet, on a recycled paper bag.

◊›◊›◊›◊›◊›◊›◊›◊›◊›◊›◊›◊›◊›◊›◊›◊›◊›

Part IV

Lamb

Chapter 12

A Revealing Look
at Lamb

selecting and recognizing quality lamb…
the primal cuts

Lamb is the meat of sheep under one year of age. After that it grows up into mutton. Mutton is more common in Europe. It can be marbled with creamy colored fat and has a distinctly stronger flavor than lamb, which makes it less popular and thus less widely available than lamb in this country. Where it is sold, mutton is considerably less expensive than lamb.

The best of the domestic lamb is readied for market when it is between five and seven months of age. This is known as "genuine spring lamb," "early lamb" or "summer lamb" and it is definitely seasonal. Genuine spring lamb can be purchased between March and September, but the lamb found on the market in the early months of spring will be the youngest and the most expensive. Imported lamb from Australia and New Zealand is available in many markets all year round. Its quality is good, but unlike American lamb, the imported variety is not graded by the U.S. Department of Agriculture and although it is produced from a very young animal, its flavor is slightly different than the lamb produced in this country.

Tiny baby lamb, bottle fed and marketed sometime between one and nine weeks of age, is renowned for its tenderness and delicate flavor. For those who can afford this so-called "hot-house lamb," it can be found at exclusive retail outlets. It is, however, strictly a delicacy.

If you are fortunate enough to have a freezer, the best time to purchase lamb is the early summer. During these months, the freshly killed lamb is still young, but not young enough to merit the expense of the first spring lamb of the season. Frequently during the summer lamb can be found on special at the supermarkets, and as such is a genuine bargain. You would be wise to stock your freezer with lamb at this time. It can be safely stored up to six months in the average home freezer.

SELECTING AND RECOGNIZING QUALITY LAMB

When selecting a cut of lamb, youthfulness is what to look for. A lean, young lamb has bright pink flesh, pure white fat, and its bones are soft, moist, and pinkish. The lean is smooth with a fine, subtle texture and firm to the touch—never flabby or tough. There is also some marbling within young lamb, but it is barely perceptible.

Another way to measure the age of lamb is by weight and size. For the consumer, this is easiest to perceive in a cut of lamb such as the leg. The highest quality USDA Choice grade leg weighs about 5 pounds and should never exceed 9 pounds in weight. Any leg heavier than 9 pounds is mature, and its meat will not be as delicately flavored, nor will it be as tender. The legs at the lowest end of the weight spectrum are the youngest and usually will be the best, as will the smaller chops, but both are often expensive and available only in the early spring. Some of these succulent young cuts are frozen by exclusive butcher shops and can be purchased at a high price all year round on special order.

Like beef, grades of lamb depend primarily on age, confirmation, and finish. The three USDA grades available to the consumer are Prime, Choice, and Good. As with beef, the Prime grade lamb is a most expensive luxury, and Choice grade the most widely available and the best buy for most consumers. All the cuts discussed in this chapter are USDA Choice grade lamb. This is the grade that will yield the delicately flavored, most tender lamb for your money.

LAMB PRIMAL CUTS

Lamb carcasses are usually purchased whole except by those outlets dealing in Kosher meat. The entire carcass weighs anywhere from 45 to 60 pounds, and is divided across the back into two equal sections: the **foresaddle** and the **hindsaddle.**

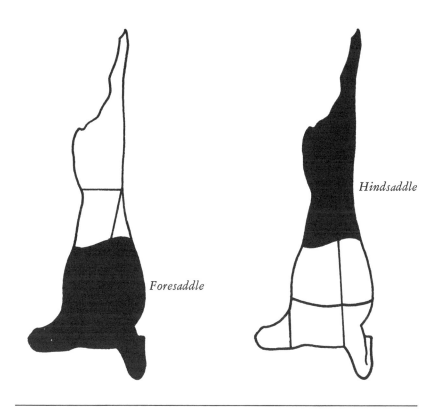

Hindsaddle

Foresaddle

The **foresaddle** represents 50 per cent of the entire lamb and encompasses four of the primal (and all of the Kosher) cuts. They are:

1. Chuck of lamb
2. Rack of lamb
3. The foreshanks
4. Breast of lamb

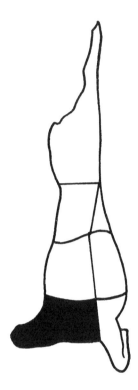

Shoulder (chuck double)

1. **Chuck of lamb** is cut from the neck, shoulder, and part of the shank portions and represents 24 per cent of the carcass. The chuck of lamb corresponds to the chuck of beef, and as such is a mobile part of the animal. Lamb is not as mature, however, at the time it is readied for market and its connective tissue not as fully developed as those in beef chuck. The meat is, therefore, still fairly tender and some of the cuts may be roasted or broiled. The chuck is usually divided into neck meat for moist cooking or grinding, and shoulder and arm lamb chops. Sometimes the bone is totally removed, and the meat is rolled into various kinds of roasts or cut into kabobs for broiling.

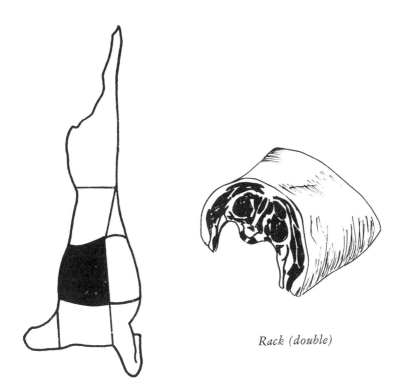

Rack (double)

2. **Rack of lamb** comes from the rib section and is about 12 per cent of the carcass. It is the most tender section of the foresaddle of lamb from which up to fourteen rib lamb chops may be cut. The rack corresponds to that section of beef that yields rib roasts and rib steaks. It is an immobile section of the lamb and has very little gristle and connecting tissue. The first three or four chops cut from the front of the rib section, that section closest to the loin, are the most desirable, as are the first cuts of beef rib. Although they have very little meat per chop, they have less gristle and fat than the end-cut rib chops, which are cut farther back, closer to the chuck. These chops are larger but have more fat and connective tissue. The meat is actually divided by large fatty streaks. The rack of lamb may also be used whole for roasting and for making those elegant, visually impressive crown roasts. The crown roast is tender and delicately flavored but provides very little meat for your money.

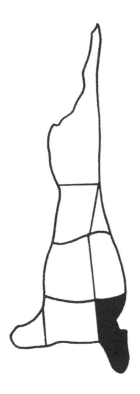

Foreshanks

3. **The foreshanks**—the two front legs of the lamb—are 4 per cent of the carcass. This portion of the lamb is more commonly referred to as lamb shanks and can provide the basis of a very inexpensive and tasty meal. Lamb shanks are fatty and contain a good deal of connective tissue but do have a fair amount of meat. They adapt best to long, slow, moist cooking but they may also be roasted. Braised lamb shanks can be simply delicious. At the table, lamb shanks will also make a more attractive and manageable dish if the butcher has divided them in half into portion-sized pieces.

Breast

4. **Breast of lamb** represents 10 per cent of the carcass and is the least meaty section of the lamb. The breast may either be cut into lamb riblets, or trimmed completely, de-boned and rolled as a roast, or ground for meatballs and casseroles. Because lamb breast is very bony, it is priced lower than the cuts from the shoulder or leg. However it may well end up costing more after it has been trimmed than those other more solid pieces of meat.

The **hindsaddle** is the rear portion of the lamb and contains two very desirable primal cuts. The hindsaddle of the lamb consists of:

1. The flank of lamb
2. Loin of lamb
3. Leg of lamb

1. **The flank of lamb** is a very stringy cut of meat, a portion of which is sometimes evident as the long "tail" on loin lamb chops. (One should be cautious when purchasing loin lamb chops not to pay for too much tail.) Because it really is not suitable for much else, most of the flank is ground into lamb patties.

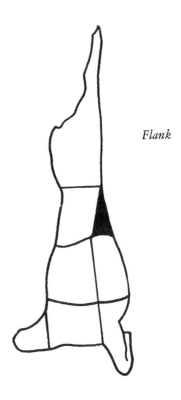

Flank

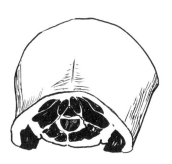

Loin (double)

2. **Loin of lamb** represents but 11 per cent of the whole carcass and is the tenderest cut of the lamb. The meat corresponds to the section of beef from which Porterhouse and filet mignons are cut. Because the loin of lamb is extremely expensive meat, it is rarely sold as a roast—although it may be. Usually the loin is cut into four or five chops that actually look like miniature versions of the Porterhouse with their own small filet sections. These can be broiled or pan broiled.

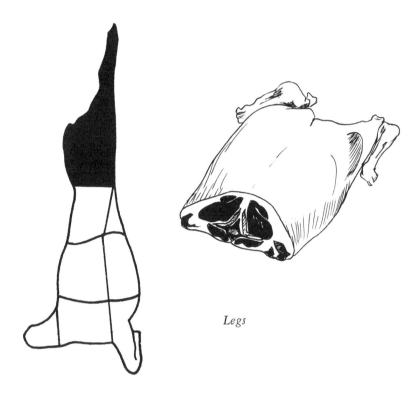

Legs

3. **The leg of lamb** is the meatiest part of the lamb, representing 30 to 35 per cent of the carcass. The leg yields the most satisfactory, easy-to-carve roasts. It may be sold whole, in sizes ranging from 5 to 9 pounds, or divided in half into two roasts: the shank end and the sirloin or rump end. Of the two, the shank is the meatier by far, the less bony and easier to carve. In some parts of the country, the leg is divided into three roasts—the center portion of the leg becomes a roast too, and the sirloin and shank sections are shortened. This practice, called an "American cut," is fairly uncommon, however. The sirloin half of the leg is also sometimes divided into three or four sirloin chops and the center of the leg cut into lamb steaks. The leg meat can also be cut into uniform chunks for kabobs. You can save pennies by purchasing a leg and cutting your own kabobs from it.

Chapter 13

Lamb—The Retail Cuts

roasts…chops…the moist-cooked meats

Compared to the other meat dealt with in this book lamb cuts are smaller and more tender. Because of the size and age of the animal, lamb is easily digested and nutritional and it is often recommended in diets for people recuperating from illnesses, as well as for small children.

The French discovered long ago that lamb is juicier and more delicately flavored when it is served very thinly sliced and pink—or even rare. This may come as a shock to those who have never had the treat, but it could also be the basis of a whole new delicious attitude. If you have never had the experience—it is certainly worth a try. Next time you roast a lamb, cook it in almost the same way you would a medium-rare roast of beef. Stop the cooking when the internal temperature registers 135° to 145° F. on a meat thermometer, rather than the conventional 175° to 180° F. Be sure to carve it very thinly. Then taste and see if you don't agree. (You may want to try this experiment first with grilled lamb chops if you feel you must accustom yourself to the experience of rare lamb.)

LAMB ROASTS

There are seven basic roasts in a whole lamb carcass. They are listed below in order of the quantity and the quality of meat each will yield for your money:

1. Whole leg of lamb
2. Shank half of leg
3. Boneless shoulder of lamb
4. Shoulder of lamb (bone in)
5. Sirloin half of leg
6. Rack of lamb
7. Loin of lamb

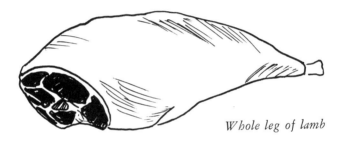

Whole leg of lamb

1. **A whole leg of lamb** of 5 to 9 pounds made up of both the sirloin or rump and the shank halfs is the meatiest and therefore the most economical of all lamb roasts. A good leg of lamb is well rounded, firm, and does not have an overabundance of exterior fat. Depending upon size, a leg will feed six to eight people generously. It is essential to purchase a leg of lamb oven ready. Oven ready means that the end of the leg, which is excess bone and contains no meat, has been removed by the butcher before the whole leg was weighed and wrapped. Otherwise you will be paying for worthless extra weight. Unfortunately when the pre-packaged legs are put on sale at a supermarket, this excess piece of bone will often be included. A leg of lamb will be easier to carve if you can have the rump bone (or aitchbone) loosened by the butcher before you prepare the meat.

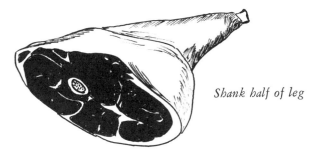

Shank half of leg

2. **The shank half of the leg** is the meatiest and easiest to carve portion of the whole leg. As with any portion of a whole roast, it does tend to lose its natural juices somewhat as it is cooked. The shank half will make a very good 2½ to 4½-pound roast—a generous meal for three to four people. As an alternative to purchasing the shank half only, you might select a whole leg and have the sirloin or rump half sliced into six lamb chops, reserving the shank half for roasting.

Boneless shoulder

3. **The boneless shoulder of lamb** is a tasty cut, but contains more fat in proportion to the lean than the leg does. Because the butcher must remove the bone, shape and tie the shoulder meat into a roast form, the cooked slices will not be as solid nor as attractive as those from the leg. The shoulder roast is less expensive, however. A 4- to 6-pound roast should serve six to eight people.

4. **Shoulder of lamb** with the bone is fatty and extremely difficult to carve because of its complicated skeletal structure. It is very inexpensive, however, and even preferred over the boneless shoulder by those

Shoulder (bone in)

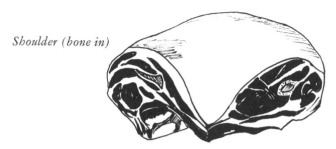

people who insist on a bone in every roast. A 6- to 8-pound shoulder will serve six to eight people. When purchasing a shoulder of lamb roast, make sure that the shoulder bones have been cracked.

Sirloin half of leg

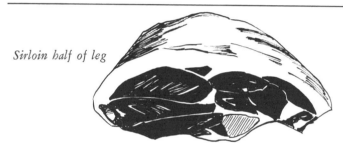

5. **The sirloin half of the leg** is usually sold for 20 to 30 per cent less than the shank half of the leg of lamb, and as such represents a real value. It does have the drawback of being extremely bony and less meaty than the shank, however. Select a large 3- to 4-pound sirloin half to feed as many people. Carving this cut can be a chore, so if you are lucky enough to have a friendly butcher, ask him to crack the bones to make carving easier.

Rack

6. **The rack of lamb** is tops for tenderness and flavor, but it is very expensive and contains very little meat. A 3-pound rack will, in fact,

contain but 1 pound of meat, and feed three people at most. The whole lamb contains two racks—one from each side of its rib cage. When these two racks are Frenched, set upright and tied back to back to form a circle, they become the elegant, even more expensive crown roast of lamb. The entire two racks combined will only yield fourteen tiny rib chop servings or just enough to provide a few tasty lamb morsels each for seven people. This roast makes an expensive, impressive indulgence—but hardly a hearty meal. When purchasing this cut, make sure the butcher cracks the chine bone between each rib to make for easy carving.

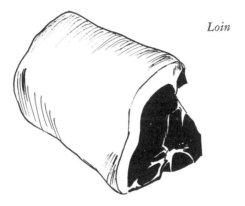

Loin

7. **The loin of lamb** is rarely seen and somewhat scarce because the loin is usually cut into loin lamb chops. As a roast, it is extremely expensive, corresponding in price to the Porterhouse steak, or when boned to a beef tenderloin roast. Even if you are willing to spend the money for a loin of lamb there is only enough meat on this roast to feed three or four people. If you purchase both sides of the loin—the saddle of lamb—you can feed twice as many people and pay twice as much. The meat on the loin roast is the tastiest of all lamb cuts, but its price practically puts it in a prohibitive bracket—except for *very* special occasions. If you do purchase a loin, make sure that very little of the flank meat (corresponding to the "tail" on a loin lamb chop) is attached. It is inferior fibrous meat and adds to the weight and the cost of the whole roast, without contributing to its overall quality.

LAMB CHOPS AND STEAKS

There are six varieties of chops and steaks in a whole lamb carcass. They will all be tender if properly cooked. Listed below, in descending order of quality, they are:

1. Loin lamb chop
2. Rib lamb chop
3. Sirloin chop
4. Blade chop
5. Arm chop
6. Lamb steak

Loin chop

1. **The loin lamb chop** corresponds to the Porterhouse steak of the beef, and indeed even looks like a miniature Porterhouse with a portion of the tenderloin filet and a comparable price tag. The loin chop comes from an immobile part of the animal and is exceedingly tender and free of connecting tissue. When choosing a prepackaged loin chop, make sure the "tail" of the chop is well trimmed. The tail consists of fibrous flank meat, and a long tail does not merit the quality to command loin chop prices. For the most delicate flavor and aroma, also make sure that the thin exterior membrane—referred to as the "fell"—is removed from the chop at the store. Most supermarkets do this automatically, but it is best to check and be certain. A chop cut ¾ to 1 inch thick may either be broiled or pan broiled. Any chop thicker than 1 inch will be better off broiled. Serve one or two chops per person, depending on the thickness of the chop. There are three to four medium-thick chops to a pound of loin.

2. **Rib lamb chop** is cut from the rack and is nearly as nice as a loin chop. It does not, however, contain the piece of tenderloin filet that a loin chop has, so it is not quite as meaty. The first three or four chops,

Loin end

Rib chops

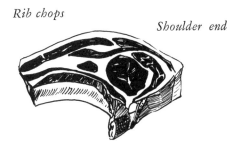

Shoulder end

those that are cut closest to the loin, are by far the most desirable. They are totally lean and devoid of interior connective tissue and can be compared to the first cut or front cut of the rib of beef. The quality of the chop diminishes the closer to the shoulder it is cut, so if possible try to purchase the first cuts. Those chops from the section nearest the shoulder —referred to as the thick end of the rib—are larger in size than the loin end cuts but less meaty because the meat of the chop is actually divided by a large interior section of fat. The closer the chop is cut to the shoulder, the larger and more developed that section will be. Rib chops cook very rapidly. Unless they are cut double thick, you will probably need to broil or pan broil two chops—¾ to 1 inch thick—per serving.

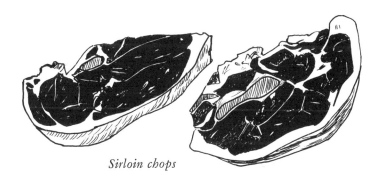

Sirloin chops

3. **Sirloin lamb chops** are sometimes·referred to as "large end of the loin" chops and are cut from the top of the sirloin or rump section of the leg of lamb, as is a sirloin beef steak. There are about four sirloin chops per whole lamb. The first two sirloin chops cut from the rump end of the sirloin contain what is called the pinbone and have more waste and bone and less meat than those chops cut farther back. The meat,

however, is just as tender and the cost is considerably less. All sirloin chops, in fact, are sold for 20 to 30 per cent less than other lamb chops because they contain a good deal of waste. A pound of lamb sirloin will only yield two of these chops. They may either be pan broiled or broiled. A chop ¾ inch thick will suffice per serving.

Blade chops

4. **A blade chop** cut from the shoulder of the lamb looks similar to a rib chop but is larger and somewhat meatier. The meat will not be as tender nor as delicately flavored as that of the rib chop, however. If you prefer your lamb chops broiled, select the blade chops from the shoulder nearest the rack, where the interior bone will be thinnest. The chops cut farther back into the shoulder will be more fibrous and less tender and must be pan fried or moist cooked. One pound of lamb shoulder will yield but two to three blade chops. A blade chop equals a serving.

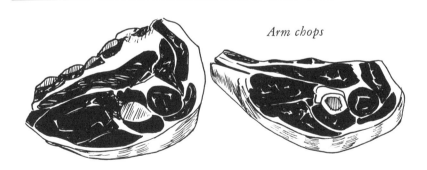

Arm chops

5. **The arm chop** cut from the shoulder contains more bone and is less delicately flavored than the blade chops. These chops cost less, however. In fact, you can save a good deal of money if you purchase an entire

square cut shoulder containing both the arm bone visible on one side and the blade bone exposed on the other. The shoulder can be divided into both arm and blade chops—about three or four of each. Thick arm chops may be broiled, thinner ones must be pan broiled or moist cooked. One arm chop is a serving.

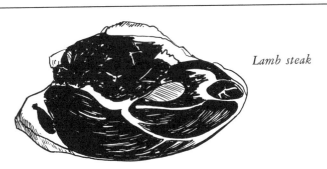

Lamb steak

6. **Lamb steaks** (or leg chops) are by far the leanest and meatiest of all the chops, but somewhat hard to come by because they are cut from the center portion of the leg of lamb which is usually sold whole or divided in half into a shank half of leg roast and a sirloin leg roast. If lean steaks are your aim, however, purchase the whole leg and have it cut into eight 1-inch thick steaks. This will be plenty for eight generous servings. If less steaks are desired, purchase only the shank half of the leg. Lamb steaks can be pan broiled, fried, or braised.

LAMB FOR BRAISING, STEWING AND MOIST COOKING

Unlike most other meat cuts that are moist cooked, lamb cuts for this purpose are generally purchased with the bone in. The reason for this is that the lamb itself is such a small animal and offers few cuts of solid meat other than those from the rack, loin, and leg. The lamb cuts normally moist cooked—the shank, neck, and breast—offer flavorful and tender meat and seem quite economical at first sight. They have about twice as much bone and waste as meat per pound, however—a point to keep in mind when figuring your budget, the number of servings required, and how much to buy.

Frequently the prepackaged lamb cubes found in the supermarket specifically labeled for stewing will contain scraps of shoulder and leg meat as a supplement to the less meaty cuts from the neck and breast. As such, these packets are quite a good value. Another way to get good value is to purchase one or more large pieces of solid meat and simply cube your own stew meat at home.

Seven cuts are listed below in order of the amount of solid lean cubed meat they offer for your lamb stew money:

1. Center-cut leg of lamb steak
2. Shank half of leg
3. Boneless half shoulder
4. Shoulder chops
5. Neck
6. Shank
7. Breast

Lamb steak (center cut)

1. **Center-cut leg of lamb steak** offers solid lean meat except for a small round bone in the center. Three steaks, each cut 1-inch thick, should yield enough meat for 1½ pounds of cubes—enough to feed four people.

2. **Shank half of leg** is a good cut for moist cooking especially during the off-season months when the meat is cut from more mature animals. Not only does the meat have a more robust flavor, but the price is right. A 4- to 5-pound shank half of leg will yield 2½ to 3½ pounds of solid meat for cubing—enough to feed six to eight people. These cubes can either be moist cooked or used for kabobs.

Shank half of leg

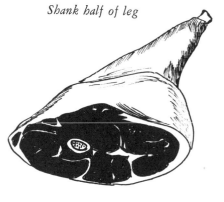

3. **A boneless half shoulder** of lamb weighing about 6 pounds will yield 3½ to 4 pounds of lean cubes or enough to feed eight people. If this inexpensive cut is available, make sure it is boneless or have the meat man at the market remove the bones for you. Otherwise, when you attempt to cube it, this cut will prove to be very unwieldy. See page 00 for illustration.

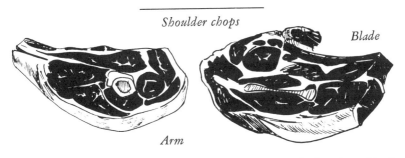

Shoulder chops

Blade

Arm

4. **Shoulder chops,** thickly cut, can be moist cooked as is, with the bone intact, or further cut into more manageable small pieces. Three thick chops will make enough stew for two—or about 1 pound of meat.

Neck

5. **Neck of lamb** is very inexpensive, but contains a good deal of bone and less meat than some of the other cuts. It will take nearly 3 pounds of flavorful lamb neck to feed three to four people.

Shank

6. **Lamb shanks** have a good deal of bone, but they are extremely inexpensive and tasty. Each lamb shank provides about enough meat for

a single person. A shank may be braised or stewed whole, but it is more attractive and easier to eat when it is split and then cooked in smaller pieces.

Breast

Riblets

7. **Breast of lamb** is a delicious budget dish, although it is extremely bony. The breast may be moist cooked whole with the fatty brisket and flank removed, or divided into tiny riblets for a most attractive stew. One whole breast will feed only two people—but it is usually a good buy.

◦)◦)◦)◦)◦)◦)◦)◦)◦)◦)◦)◦)◦)◦)◦)◦)◦)

Part V

Veal

Chapter 14

❖❖❖❖

Veal—The Status Meal

*origins...grades...recognizing quality
...the primal cuts*

Veal is young beef—less than a year old—and is produced exclusively from dairy cows, such as the Holsteins. The youngest of all veal —bob or bobby veal—is only one to five days old when it is slaughtered and de-boned for meat. Because the veal is so small it is unsuitable for retail cuts, so almost all of this veal goes to processors of canned, frozen, and convenience foods where it is made into breaded cutlets and patties. Because it is so tender, bob veal is also widely imported and used as the basis for baby foods.

The vealer which produces the best veal is usually readied for market when it is three months old and weighs anywhere from 100 to 150 pounds. Any older vealer weighing more than 225 to 250 pounds is known as a calf. The meat of a calf is coarser in texture and has a dense outer layer of fat that the vealer doesn't have. Rather than sell it as veal, some retail stores around the country offer it as baby beef. It has, however, neither the full flavor of beef nor the delicate flavor of veal.

The amount of veal we eat every year is rapidly diminishing—down from an average of 12 pounds a year per capita in the mid-forties to less than 3 pounds a year per capita today. This decline in consumption has little to do with the discrimination of American palates but rather with the relative shortage of good veal and the fact that it is quite expensive.

Raising veal is indeed an expensive operation. The young animals are pampered and coddled from the day they are born and many are completely milk-fed. In years past, the young animals fed on their mother's milk, but technology has developed a better, non-fat milk substitute utilizing the natural by-products of the butter and cheese industries.

The short supply is due to the fact that the numbers of dairy cattle in this country are dwindling. Dairy cows are now simply more efficient, and because of the sharp rise in milk production per animal, it takes fewer dairy cows to produce the milk we need and its related products. Thus fewer baby calves are being born from dairy stock, and that means less veal at your meat market.

Veal is graded by the United States Department of Agriculture in the same way that beef and lamb are: confirmation, finish, and quality. The USDA grades available to the consumer are Prime, Choice, and Good. A trademarked name for veal that is widely used by one of the country's biggest purveyors to describe the top of his line is "plume de veau." Those retail cuts touted as plume de veau are, indeed, some of the finest and most expensive available. They should not, however, be confused with the USDA standards: Prime, Choice, and Good. All the veal discussed in this chapter is USDA Choice grade.

In choosing quality veal the most important factor to consider is its color. The high quality veal has exterior fat that is clear, firm, and white with never a hint of yellow discoloration. The lean is a very pale pink, almost grayish-white, with no marbling as veal contains very little fat. The texture of the meat is fine, fairly firm, and velvety in appearance. The bones of young veal are soft, moist, and narrow and have a reddish tinge to them. Lesser quality or older veal will have darker pink or nearly red meat and white, more developed bones.

The keeping qualities of veal are not as good as those of other meats due to its skimpy outer layer of fat. Freshness can be determined by the smooth, moist texture of the flesh as opposed to the wet and slimy lean of veal that has been around a long time. Another guide to fresh veal is that the meat near the bone is firm, never soft or flaccid to the touch.

There is also a certain seasonal aspect to purchasing veal. Veal tends to be more prevalent during the winter months. This is due to the fact that Holstein vealers are less adaptable to surviving in cold weather than baby beef cattle, and some of those dairy/veal states can get very chilly indeed. If you are willing to pay the price, veal is available year round in most urban areas at specialized meat outlets.

THE PRIMAL CUTS

Although the bone structure of a veal calf corresponds exactly to that of beef, the initial method of dividing it for retailing is slightly different. Instead of splitting it into sides like beef, a veal carcass is merely halved into a **foresaddle** and a **hindsaddle.** All the Kosher veal is cut from the foresaddle. The hindsaddle, however, yields the majority of the better cuts and commands a higher price at both the wholesale and retail levels.

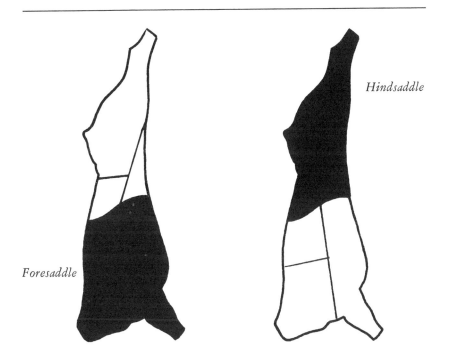

Hindsaddle

Foresaddle

The **foresaddle** and the **hindsaddle** of veal are further subdivided into six primal cuts. They are:

1. Shoulder (chuck)
2. Rack of veal
3. Breast and foreshank
4. Legs of veal and hindshank
5. Loin of veal
6. Flank of veal

The three **foresaddle** primal cuts are:

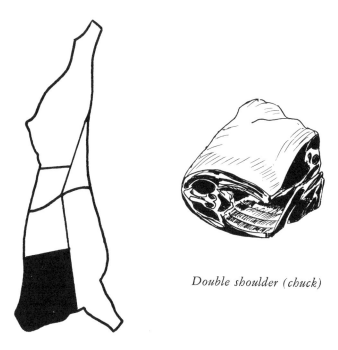

Double shoulder (chuck)

1. **The shoulder** (**chuck**) of veal, represents 26 per cent of the entire animal. Like the chuck of beef, the veal shoulder is a mobile or carrying muscle, thus a less tender cut. The shoulder may sometimes be cut into roasts with the bone intact—but more often is de-boned and tied into roasts, cut into veal steaks, chops, shoulder cutlets, or stew meat. The larger cuts may be roasted with a little extra fat added to the exterior surface, but they will be more tender if braised or moist cooked. The smaller cuts from the shoulder—the steaks, chops, and shoulder cutlets—may either be moist cooked, pan fried, or sautéed.

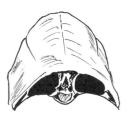

Rack (double)

2. **The rack of veal** represents only 9 per cent of the animal and is the most desirable section of the foresaddle. The rack, which corresponds to the rib section of beef, is usually cut into rib chops or sliced into ultra-thin veal scallops for sautéeing or braising. It may be purchased whole on special order with its bones for roasting or braising. This cut can also be transformed into an elegant, expensive crown roast of veal, or completely de-boned for a roast cut.

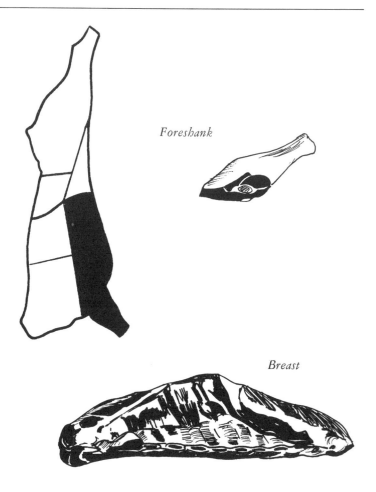

Foreshank

Breast

3. **The breast and foreshank** represents 10 per cent of the total animal and yields the least desirable meat of the foresaddle and also the lowest price. The breast may be braised whole either boneless or with the bones and with or without a pocket inserted for stuffing. The breast may also be cut into riblets for moist cooking. Another method of cooking—roasting—can best be accomplished with a breast that is completely de-boned, then stuffed, rolled, and tied. When one purchases the whole foreshank it is usually divided into portion-sized pieces and must be moist cooked for maximum tenderness. The foreshank is the basis for that famous Italian dish: Osso Bucco.

The three **hindsaddle** primal cuts:

Legs and hindshanks

4. **The legs of veal and hindshanks** are the largest of the primal cuts representing 39 per cent of the entire veal carcass. Each leg may either be de-boned, rolled, and tied for easy carving or sold with the bone intact. The leg is usually subdivided into various roasts for consumers. It may be divided in half, like a ham with the bone intact, into a rump end and a shank end. Or boned out completely, it divides into roasts of varying quality. The roasts from the shank and the butt may be roasted or braised. The hindshank must be moist cooked, and the smaller cuts—such as scallops and leg cutlets—may either be braised or sautéed. The whole leg, with the bone intact, is so large that it is usually only sold to restaurants which cater to crowds.

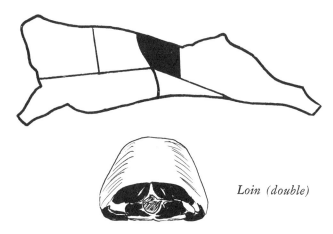

Loin (double)

5. **The loin of veal** corresponds to that section of the beef from which all the sirloin steaks and filet mignons are cut. As the loin represents only 7 per cent of the entire veal, it is indeed an expensive, elegant cut. Most frequently it is cut into loin chops for braising or sautéeing. The loin of veal may also be sold with the bones as a roast, or less often as a boneless, rolled roast.

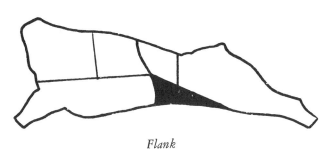

Flank

6. **The flank of veal** represents only 2 per cent of the carcass. Cuts from the flank are the least desirable cuts of the hindsaddle. Most of the meat, therefore, is put through a machine and cubed to break down the connective tissue, then sold as cutlets. The flank meat may also be ground for making meat loaves and veal patties.

Chapter 15

Versatile Veal— The Retail Cuts

cooking methods...cutlets...chops, steaks, and roasts

As akin in origin as beef is to veal, when it comes to flavor and cooking methods veal is a whole different animal. Veal has very little fat, and no marbling, and it also is the most delicately flavored of all meats. These facts can be attributed to both the age of the animal at the time it is readied for market and the bland milk or no-fat milk substitute diet it has been fed. One of the other characteristics that differentiate veal from beef is the way in which the entire carcass is cut.

The greatest percentage by far of veal available to the consumer today is in the form of cutlets. A cutlet is a small portion of solid, lean meat with no splits—no muscles divided by gristle. Many cutlets of veal are processed by manufacturers—breaded, sauced, and frozen into convenience foods.

A derivative of the cutlet is the ultra-thinly sliced, flat veal scallop. Veal scallops are what the Italians refer to as "scaloppini" and the French call "escalopes de veau." In any language they are small slices of veal which may be floured or breaded, stuffed, rolled and braised, and also may be pan fried or sautéed *au naturel* in butter or other shortening.

Cutlets and scallops can be sliced from four of the primal cuts: the rack (rib), the shoulder (chuck), the loin, and the leg. When purchasing prepackaged cutlets or scallops at a supermarket, ask for their origin or look for the primal cut on the label. Those cut from the loin and the rib will be the most expensive and the most tender because they come from an immobile part of the vealer. Those from the shoulder will be the least expensive and the least tender.

An expeditious way to tenderize veal scallops prior to cooking them is to place each scallop between two pieces of waxed paper, then pound the meat with the side of the cleaver, wooden mallet, or other flat utensil. The pounding helps to make the scallop flatter and thinner—desirable qualities for a scallop. Pound with care, however, so that you will not break or tear the veal.

Contrary to what many cookbooks recommend, broiling should never be attempted with veal. The exception is high-grade loin or rib chops from more mature, fatter calves. Otherwise, broiled veal will become dry, brittle, and tasteless.

The best methods of cooking veal are roasting, braising, sautéeing, and stewing. The larger, more expensive cuts can be dry roasted with some extra fat or barding added to their exterior surface (see pages 00 and 00). But they tend to take better to moist cooking. Regardless of the method of cooking, veal should always be cooked well done.

A stew of veal can also provide a culinary delight. Select cuts for stewing from the neck, shoulder, heel, shank of the leg,

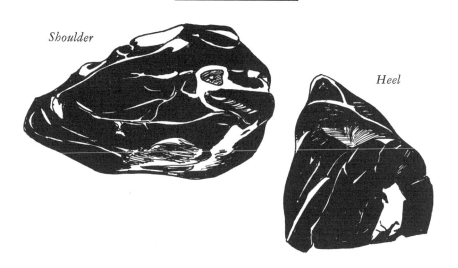

Shoulder

Heel

Shank (cross cuts)

breast, and short ribs of veal. Allow about ⅓ of a pound of boneless meat per serving—or up to ¾ of a pound if the meat is bony.

One of the beauties of veal is its infinite capability to adapt to recipes. Versatile veal is to that extent somewhat like chicken. Scallops for instance, may be stuffed with infinite combinations of ingredients for veal birds, and veal cutlets may be cooked in countless ways. The stew cuts and the roasts capture the essence of the ingredients with which they are cooked as well as imparting their own delicate flavor. In fact, if veal were more widely available and less costly someone would be bound to write a book on 1,001 ways to prepare veal. Until that day, content yourself with the hundreds of excellent existing recipes and your own ingenuity.

CUTLETS, CHOPS, STEAKS, AND RIBLETS

Most of the veal purchased today is in the form of cutlets, chops, and steaks. Those cuts that are the most desirable and frequently available are listed below in the order of the amount of meat they will provide for your money. Tenderness is of secondary importance because most of this veal is prepared by one of the moist cooking methods or sautéed. One word of caution: some of these cuts can be very expensive—at this writing upwards of $3 per pound.

1. Veal leg cutlet
2. Veal loin chop
3. Veal rib chop
4. Veal arm steak
5. Veal blade steak
6. Veal riblets

Leg slice   *Cutlet*
                                                  ↓

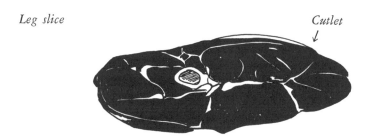

1. **The veal leg cutlet** tops this list for practical purposes. Not only is it the most widely available cut of veal, sliced from the largest primal cut, but also it is all meat—no fat, no bone, no waste. The leg cutlet is usually purchased in individual-sized portions of about 1½ to 2 ounces each, varying from ¼ to ½ of an inch in thickness. When purchasing cutlets to cook at the same time, make sure all are of equal thickness so that they will cook evenly. Depending upon appetites, you will need two to three cutlets per serving. The leg cutlet may also be further subdivided into scallops at less cost than scallops from the loin or rib sections.

Loin chops

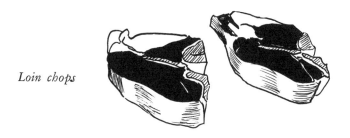

2. **The veal loin chop** is a meaty chop which contains part of the tenderloin. It also contains a good deal of bone and resembles a mini-Porterhouse steak. Veal loin chops are usually cut ¾ to 1 inch thick. One loin chop provides an ample serving. When the loin chop is sold with a portion of the veal kidney encased in its center, it is known as a kidney chop.

3. **The veal rib chop** is less meaty than the veal loin chop and has none of the fine meat of the tenderloin. But it is comparable in price to the loin chop. Rib chops cut from the section of the rack nearest the

Rib chops

loin are the best—those cut nearer the shoulder, more bony and fatty. Depending on the thickness of the chop one or two chops will be needed per serving.

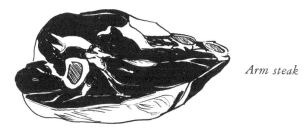

Arm steak

4. **The veal arm steak** is cut from a section of the shoulder close to the rack. It costs considerably less than the cuts described previously and the meat is of lesser quality because it is from a mobile part of the vealer. The arm steak contains a small section of bone and some connective tissue. One arm steak, ¾ to 1 inch thick, will provide enough meat for two people.

Blade steak

5. **The veal blade steak** is cut farther back on the shoulder and contains more bone and less lean meat than the arm steak. A blade steak is one of the least costly cuts of veal you can buy. It can be cut into

cubes for stewing or cooked as is. One blade steak ¾ to 1 inch thick will be enough to feed two people. Cubed at home and combined with other ingredients, blade steak will provide enough meat to make a stew for three.

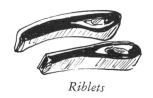

Riblets

6. **Veal riblets** are cuts from the breast of veal—a real veal bargain. Each riblet contains a piece of the breastbone. The meat, which is moist cooked, is delicately flavored—but there is very little of it. In fact, it will take all the riblets from an entire breast of veal to provide enough meat to serve four people.

VEAL ROASTS

Veal roasts can be cut in all shapes and sizes from the better primal cuts of the veal such as the leg, loin, and rack, and from some of the lesser primal cuts such as the shoulder and breast.

Veal roasts, by and large, are de-boned, rolled, and tied before they reach the supermarket meat section. They can be dry roasted with fat added to their exterior surface, or even better, braised in a liquid: stock, or wine, flavored and scented with spices and herbs. Listed below in descending order of desirability—the amount of meat you are likely to get for your money—are six veal roasts. When selecting a roast of veal that has been de-boned, rolled, and tied, always look for one that is uniform in shape from end to end so that it will cook evenly.

1. Whole shank half of leg
2. Rump half of leg
3. Shoulder of veal
4. Loin of veal
5. Whole breast of veal
6. Rib roast

Shank half

1. **A whole shank half of leg** with the bone intact provides the most meat, least waste, and greatest ease in carving. A whole shank half of leg weighs from 5 to 7 pounds and will provide eight to ten servings.

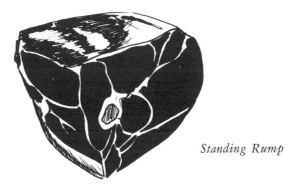

Standing Rump

2. **The rump half of leg** is either sold with the bone intact, or de-boned, rolled, and tied into roasts of varying sizes and shapes. The meat is of very high quality, but like the sirloin half of a leg of lamb or the butt end of a ham, it has a complicated bone structure and with the bone intact is harder to carve than the shank half of the leg. With the bone it is known as a standing rump roast. The whole rump end of leg with the bone weighs 5 to 7 pounds and will serve seven to eight people. A pound of boneless rump will serve three or four.

3. **Shoulder of veal** is always de-boned, rolled, and tied into roasts of various sizes. The meat is not of the quality of that of either of the roasts from the leg, but it costs less and can be just as tender if roasted

Shoulder

with some extra fat added to its exterior. The shoulder roasts are available in sizes ranging from 4 to 7 pounds. One quarter to ⅓ pound of boneless meat will provide a serving.

Loin

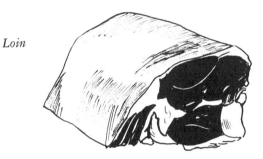

4. **Loin of veal** is the most expensive and desirable cut of the entire veal. It corresponds to the section of beef from which the Porterhouse, T-bone, and sirloin steaks are cut—an immobile part of the animal. It is listed here as selection #4 because when it is divided into roasts it is so expensive that it is in the prohibitive bracket for most consumers. The actual amount of meat compared to the amount of bone is low. In that sense the loin of veal compares to the loin of pork, but it is much more expensive. It ranges in size from 4 to 5 pounds, and will feed about the same amount of people. Since the loin is usually cut into chops, you will have to order this cut in advance.

5. **A whole breast of veal** is a low-cost cut as veal roasts go. The whole breast is extremely bony, however, and contains little meat. In order to make it stretch further, a pocket is often cut into the breast so that it may be stuffed. For a do-it-yourself double veal meal just follow the directions in Chapter 20, "The Step-by-Step Guide to Cutting Meat."

Breast

A whole breast of veal roast weighs 5 to 9 pounds or more and will feed about the same amount of people without stuffing. The breast may also be completely de-boned, rolled, and tied.

Rib roast

6. **The rib roast of veal** is really a rarity so it ranks last on the list. This roast has top-quality meat but lots of bone. It may be sold whole or divided into smaller roasts with the bone intact, or de-boned, rolled, and tied. The ordinary rib roast ranges in size from 3 to 5 pounds and only provides enough meat to feed three or four people. A regal version of the rib roast is the crown roast of veal. The rib roast is most frequently cut into chops, but if you place a special order in advance you might just get one.

◊+◊+◊+◊+◊+◊+◊+◊+◊+◊+◊+◊+◊+◊+◊+

Variety Meats
and Sausages

Chapter 16

❖❖❖❖❖

Variety Is...a Bargain!

liver, tongue, sweetbreads, heart, kidney, brains,
tripe, and oxtail

The term "variety meats" was probably created by an enterprising butcher to help overcome the stigma attached to eating offal, innards, organs, or glands. Regardless of their current designation, variety meats have long been held in high esteem by people of adventuresome palate and by nutritionists as well. They offer the sophisticated shopper not only some of the greatest amounts of protein available per food dollar, but also a rich source of vitamins and iron. And, except for a thin outer membrane or minimal amounts of fat, variety meats are practically wasteless. For all but the most finicky eaters, variety meats can be the basis of delicious, highly nutritious, budget-minded meals.

One word of caution: all variety meats are extremely perishable and should either be frozen when they are purchased or eaten within a day or two.

LIVER

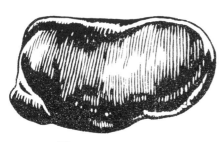

Liver
American Meat Institute

Liver is the world's single richest source of iron for the diet. Of all the variety meats, it remains the most popular with consumers. The highest quality liver, therefore, is somewhat hard to come by because the demand exceeds the supply. Livers are generally graded according to size. The livers from the younger and lighter animals are usually the more desirable.

Liver is usually sold in slices and is highly perishable, so it should be eaten within a day or two after it is purchased. Otherwise it may be frozen, but be sure to check the label or ask the meat man to make certain that it hasn't been frozen and defrosted before. If it is refrozen, liver will suffer a loss of quality. Like all organ meats, it is receptive to the rapid growth of bacteria, and refreezing will increase the bacterial count. All liver should be purchased in ultra-thin slices of uniform thickness. To a certain extent its color will be determined by the breed and feed of the animal from which it came, but that color should be lively and have a "bright bloom," never slimy or dull.

If it is still intact, the thin, opaque, outer membrane should be removed before the liver is cooked. For maximum tenderness and vitamins, pan frying or sautéeing is the cooking method most highly recommended for liver. If properly prepared, liver should be moist and fork-tender, not dry and tough, as so often happens if it has been overcooked. Whether pan frying, sautéeing, or broiling, liver will require about one minute of cooking time per side, and no more than two minutes. One pound of liver will provide four average servings. Listed below in order of quality, tenderness, and flavor are the four kinds of liver:

1. **Veal (and/or young calves') liver** is the most expensive, but also the most tender and delicately flavored. A light reddish brown color is the sign of quality. Liver from older cattle will be a darker reddish brown. Sometimes beef liver is labeled and sold as calves' liver—at calves' liver prices. Color can provide the clue for the observant shopper. A whole veal liver usually averages 2 to 4 pounds, a calf's liver 7 pounds or slightly more. Since liver is generally purchased sliced, one quarter

pound may be figured as the size of an average serving. Calves' liver and veal liver may be broiled, lightly braised, pan fried, or sautéed. Care should be taken not to overcook.

2. **Beef liver** has a dark reddish brown color and a stronger flavor and aroma than calves' liver. It is also less tender but less costly—a good choice for budget-minded cooks. Good quality whole beef liver weighs from 8 to 12 pounds. A whole liver of lesser quality and from a more mature animal can weigh up to 25 pounds and is dark in color. Beware of the store that features both "beef liver" and "select beef liver." The "beef" liver is more likely to actually be cow's liver—the product of an animal six or seven years old. The danger here lies in the fact that most cows have been dosed with medicines at one time or another —and those drugs have a tendency to linger in the liver. Like calves' liver, beef liver may be braised, pan fried, broiled, or sautéed and should not be overcooked. Thanks to modern processing, much of the beef liver available today has been skinned and deveined before it reaches the market. If the skin is still intact, however, it should be removed prior to cooking, because it shrivels up and causes the meat to curl. You will need one quarter pound per portion.

3. **Lamb liver** is more tender and less expensive than beef liver. Because of its strong resemblance to calves' liver, which would be difficult for even the sharpest shopper to discover, some unscrupulous butchers have in the past tried to palm it off as such at calves' liver prices. One distinguishing difference from calves' liver is a sharp, distinctive odor. The average whole lamb liver weighs only 2 pounds, so the lamb liver slices will have a smaller circumference. Like all liver, it may be broiled, braised, pan fried, or sautéed. You should figure one quarter pound per serving.

4. **Pork liver** is primarily used for liverwursts and commercial patés and is not widely available to the consumer. When it is available, it is very inexpensive and just as nutritious as other livers. It does have a very strong odor and flavor, however, which may not suit everyone's taste. A whole pork liver weighs 3 pounds or more. Either broil, braise, pan fry, or sauté one quarter pound, thinly sliced, per serving.

TONGUE

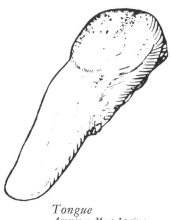

Tongue
American Meat Institute

Beef and veal tongues, which weigh on the average 3¾ and 1½ pounds respectively with the bone in, may be purchased either fresh, corned, pickled, or smoked. The best tongues—either fresh or cured—weigh under 3 pounds and have a finely knit texture. Pork and lamb tongues are most often sold canned and ready to eat. One expert we consulted definitely felt that the quality of cured beef and veal tongues was the most satisfactory.

Tongue is distinctively flavored. It has a thick outer layer of skin which means that long, slow, moist cooking is a must to make it tender enough to eat. The general method of cooking fresh tongue is simple: the whole tongue is scrubbed, then simmered in water to cover with seasonings added if desired, and depending on size, cooked for two to four hours until tender. Consult your cookbook about the seasonings. After it is fully cooked, the tongue is quickly plunged in cold water. The skin is split and peeled off—like a glove. The processed tongues—those that are pickled, smoked, or corned—may have to be soaked in water before they are cooked. Be sure to check the label.

Before serving, cut away the roots, small bones, and gristle. Carving is done on a slight angle. Tongue is served thinly sliced, either hot or cold, and usually accompanied by spicy mustard or horseradish sauces. (Cold tongue will be more juicy if it is allowed to cool in its cooking water.)

A whole 3-pound beef tongue will provide enough meat for a buffet of ten or more people. A whole 1½-pound veal tongue will be enough for five or more dinner servings. One quarter pound of tongue is an adequate serving.

SWEETBREADS

Sweetbreads
American Meat Institute

Sweetbreads along with brains are the most delicately flavored variety meats. Sweetbreads are the small thymus glands from the neck and heart of young steers, calves, and lamb. As the animal matures, these glands shrivel and disappear, so they are never found in older animals.

Good sweetbreads should be light, bright, and rosy in color. The larger they are in size, the more desirable they are to discriminating consumers, and the higher the price they can command. Size does not affect the flavor, however. As a result, lamb sweetbreads, which are tiny, have very little consumer appeal and are rarely sold on the market. Veal sweetbreads are the most sought after.

Sweetbreads are generally packed and sold in sets of two, or pairs— the round lobe from the heart, the elongated lobe from the neck. These paired sweetbreads are generally sold as a gourmet item, so single unmatched sweetbreads may sometimes be purchased at a considerable discount.

Sweetbreads are washed and soaked before they are cooked. The outer membrane is removed either before or after cooking. Although it is not entirely necessary, they are usually precooked in simmering water with a little lemon juice or vinegar added for about fifteen minutes to help firm their texture. Precooking is always a good idea if you do not plan to use them right away because sweetbreads are so perishable in their raw state.

Sweetbreads may be braised, fried, sautéed, or broiled. You will need one quarter pound per serving, or less if you plan to cook them with other ingredients.

HEART

The heart is a very active muscle, by definition tough and

Heart
American Meat Institute

elastic. Veal, lamb, pork, and beef hearts are no exception. Heart is the least expensive of all the variety meats and offers the bonus of rich nutritional benefits. It is also very tasty. The effort involved in its preparation is definitely worthwhile.

Veal and lamb hearts are the most satisfactory, and can be cooked a number of ways. They may be roasted with added fat, plain or stuffed. Or they may be braised, or simmered in liquid to cover until they are tender.

Beef heart is the least tender and must be cooked longer—up to four hours or more. It offers less flexibility in the cooking methods as it must always be moist cooked.

All hearts should be thoroughly washed before cooking and the membrane inside that divides the two heart chambers should be removed—particularly if the heart is to be stuffed. A whole lamb heart weighs one quarter pound and provides one ample serving. Veal and pork hearts are larger—one-half pound or more—and they will serve two or three. The beef heart weighs about 4 pounds and will serve a dozen or more people if it is stuffed.

KIDNEYS

Kidneys
American Meat Institute

Kidneys are highly nutritious, distinctively flavored, and considered a great delicacy by their fans. Fancy "English chops" are those veal and lamb loin chops cut with a portion of the reddish-brown kidney attached. The kidneys of veal, lamb, and pork are the most tender. Veal ranks highest for flavor. Most pork kidneys are used for patés and terraines. Beef kidneys have a more pronounced flavor and strong odor and are less acceptable because they require long moist cooking to make them tender enough to eat.

Veal, lamb, and pork kidneys may be cooked whole or sliced. They may be pan broiled, sautéed, cooked in stews, baked in pies, served sauced or plain with a minimum of effort (and delicious results).

Like liver, kidneys cook rapidly and caution should be taken not to overcook them. Kidneys, too, are perishable and should be prepared promptly after they are purchased. Before cooking, cut away any outer fat, remove the outer membrane and the hard inner core. Wash the kidneys well in cold water and dry them with paper towels.

Depending on the ingredients with which they are cooked, a whole beef kidney will feed four or more people, a veal kidney will suffice for three or more, a pork kidney provides a portion, and two lamb kidneys will be necessary per serving.

BRAINS

Brains
American Meat Institute

Although somewhat hard to come by, beef, pork, veal (calf), and lamb brains are all exceedingly tender and very delicately flavored. The most highly preferred brains—those for which Europeans and gourmets salivate —are the calves' brains. Choose brains that are a clean, light pink color and free from blood clots and stains. Beef brains are the least desirable.

Like the other variety meats, brains are very perishable and should be promptly precooked. Precooking is, in fact, a prerequisite to most methods of preparation and it also enhances the keeping quality of brains. Before precooking, soak the brains in cold water and remove the outer membrane. Then simmer them for about twenty minutes in salted water, adding a tablespoon or two of lemon juice or vinegar, and other seasonings if you desire. Another method is to simmer them in milk. This step will firm their mushy consistency for use in other recipes.

Brains may be sautéed or cut in small pieces and fried, creamed, or scrambled with eggs. Depending on the other ingredients called for in your recipe, 1 pound of brains will serve four people.

TRIPE

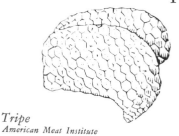

Tripe
American Meat Institute

Tripe comes from the first and the second stomachs of beef. It may be purchased fresh, pickled, corned, or fully cooked in one of two styles: plain or honeycomb. Of the two, honeycomb is the more preferable.

Tripe is not as highly nutritious as some of the other variety meats, but it is relatively inexpensive and delicately flavored. All tripe is chemically washed, soaked, scraped, and scalded to remove the inner stomach wall, and precooked before leaving the packing house. Tripe is, however, very tough, and will require up to two more hours of precooking at home to "tenderize" it—unless, of course, it is purchased fully cooked. After precooking, tripe is most often cut into pieces and broiled, fried, braised or used as the secret ingredient in the famous "pepper pot" soup. If served plain, 1 pound of tripe will serve four.

OXTAIL

Oxtails

Oxtails are not the tails of ox at all, but rather the tails of beef cattle. They contain a considerable amount of meat which has a rich, hearty flavor. They are, however, extremely tough, contain a good deal of bone, and require long, slow moist cooking.

Oxtails are particularly adaptable to soups and stews and are usually cut into joints prior to cooking. Two tails in combination with other ingredients will provide enough meat for a stew large enough to feed five or six people.

Chapter 17

❂❀❂❀

Sausage and Its Sidekicks–Lunch Meat and Cold Cuts

how meat is processed... variety... glossary

It is estimated that there are more than two hundred types of sausage and its related products produced today. Many have endless variations. Sausage is, in fact, one of the oldest processed foods. The Mediterranean peoples were grinding it out more than 1000 years B.C. The European sausage makers closely kept the secrets of their local specialties. When the immigrants came to this country, however, they brought these secrets with them and formed the basis of the modern American sausage industry. Today there are more than 4,000 U.S. sausage processors producing more than 5 billion pounds of sausage a year.

Sausage is shaped into links, rings, and rolls, formed into patties and loaves, sold sliced or in bulk form. It consists of one or more kinds of chopped or emulsified meat which may either be stuffed into natural casings of animal entrails, or more commonly one of synthetic materials. "Skinless" sausages have this casing removed by machine at the processing factory.

The primary ingredients of sausage and cold cuts is either pork, beef, veal, or variety meats, or any combination of these meats. The meat is finely ground, coarsely chopped, or emulsified into a smooth, almost paste-like consistency—like frankfurters. In most instances the meat is mixed with other ingredients such as dry milk, eggs, salt, pepper, gelatin, corn syrup, cereal, nuts, onion, garlic, and stabilizers.

The distinctive flavor characteristic of each type of sausage is further defined by the addition of one or a number of herbs and spices. Forty-five different kinds have been known to be used at one time or another. They include allspice, anise, caraway, cardamom, cassia, celery seed, cinnamon, cloves, coriander, chives, cumin, dill, fennel, bay leaves, ginger, mace, marjoram, mustard, nutmeg, paprika, parsley, sage and thyme. One recipe for frankfurter includes white pepper, cloves, coriander, nutmeg, and sugar.

A recent controversy has been raging between consumer advocates and the pork industry over the inclusion of chemical salts—nitrites and nitrates—which act as preservatives, kill botulism, and cause certain color changes to occur in the sausage formula. The pork people claim that the aesthetics of their product would be destroyed without these additives. (Frankfurters would turn out dark grayish-red.) Consumer groups argue—but have yet to prove conclusively—that nitrites and nitrates could cause cancer. As of this writing, the controversy has yet to be settled.

In a typical modern supermarket, the selection of sausage and cold cuts is vast, even overwhelming: many offer the consumer a choice of as many as 150 brands, types, and varieties. In an attempt to cut through this possible confusion, five basic classifications of sausage are sorted out below:

Fresh sausage
American Meat Institute

1. **Fresh sausage** is composed of uncured raw meat. Bratwurst, bockwurst, and Italian sausage, as well as many of the mild-flavored breakfast-type pork sausages fall into this category. The fresh sausages are extremely perishable and must be thoroughly cooked and eaten within a day or two after they are purchased. They may be simmered—or they may be broiled, grilled, pan fried, or pan broiled. If you would like to lower some of the fat content as they cook, puncture the casing with a fork to allow the fat to drain out.

2. **Smoked sausage** is composed of cured, smoked meat. There are two types: uncooked and cooked. Smoked country-style pork sausage and mettwurst are examples of uncooked smoked sausage. Frankfurters and bologna are examples of cooked. Uncooked smoked sausage must be thoroughly cooked and eaten within a

Smoked sausage
American Meat Institute

day or two after it is purchased. Any of the cooking methods recommended for fresh sausage will do. Cooked smoked sausages may be eaten as is, but they are usually heated and eaten warm. Well wrapped, cooked smoked sausage will last up to two weeks in the refrigerator.

Cooked sausage
American Meat Institute

3. **Cooked sausage** can be distinguished from the preceding classification in that it consists primarily of uncured meats. This category, which encompasses most of the liver sausages and Braunschweiger, is purchased ready-to-eat and generally served cold. If left in its original wrappings, it will last up to a week in the refrigerator.

4. **Semi-dry and dry sausages** are cured, sometimes cooked, and then air dried to remove moisture. Some of the hard-dried type are aired for a period of one to six months. The texture of the semi-dry type is somewhat softer than that of the firm, hard-

Semi-dry and dry sausages
American Meat Institute

dried sausages. Most of the salamis and cervelats are of this category. They do not require cooking, are ready-to-eat, and will stay well in the refrigerator for two weeks. Refrigeration is necessary if you plan to keep them any length of time, but they make ideal picnic items because they tend to keep well for short periods of time even without refrigeration.

Cooked meat specialties
American Meat Institute

5. Cooked meat specialties are not technically sausages but related, ready-to-eat, processed meat products including cold cuts, lunch meats, sandwich spreads, luncheon loaves, dried chipped beef, and jellied meats. These prepared meat products are popular for sandwiches, salads, and snacks. They all require refrigeration unless they are canned and will stay fresh for about a week.

Variety and versatility are characteristic of sausage and its related products. They may be eaten for breakfast, lunch, dinner, snacks, and canapés. Generally a pound of sausage in combination with other foods will be enough to serve four. For breakfast you will need only about 3 ounces per portion. The thinly sliced lunch meats can, of course, be stretched further on sandwiches. Frankfurters are generally packaged ten to the pound. An adult will generally eat two per meal—but for picnics and barbecues you may need to estimate more.

The frankfurter is, of course, far and away the most popular of all sausages. It is estimated that the average American eats at least eighty frankfurters a year. According to the standards set up by the United States Department of Agriculture, a frankfurter can contain up to 30 per cent fat. In recent years consumer advocates have pleaded for a reduction in that allowable amount of fat, and, in fact, the average frank actually does contain slightly less. The U.S. Department of Agriculture also allows the frank recipe to include up to 10 per cent water, and up to 3.5 per cent fillers and non-fat dry milk (unless it is labeled "all meat" or "all beef").

The exact ingredients of frankfurters will differ from brand to brand. Like all packaged sausage products, the frankfurter label lists those ingredients in descending order according to weight. The major ingredient will be listed first, the least significant last, and the other ingredients will be listed in order of their importance. One recent survey at a supermarket proves why it pays to read the label. The observer noted that a particular brand of "all meat" frankfurters listed its ingredients as: beef, water, pork, seasonings, and stabilizers. A consumer who opts for this particular brands ends up paying for more water than pork.

All frankfurters will be labeled one of three ways:

"All meat"—These frankfurters are a combination of beef, pork, water, and spices, but contain no fillers such as dry milk or soy flour.

"All beef"—This variety contains only beef, water, spices, and stabilizers but no fillers. The garlic-flavored Kosher frank is in this family.

"Frankfurter"—This type contains up to 3.5 per cent dry milk and soy flour as well as both pork and beef, spices and water.

Of these three kinds, the "all meat" frankfurter is the most popular, the "all beef" the most expensive. Ironically, the plain "frankfurter"— the type that contains the fillers—provides the most protein because of those fillers and is the least expensive of the three.

To help you sample your way through the vast array of sausages, we have included at the end of this chapter an alphabetical list describing what's generally available. A few helpful tips in selecting and shopping for sausage are:

· Try to avoid rumpled packages of sausages or cold cuts, or those that have accumulated moisture inside of the package. They will not, as a rule, be of the freshest quality. Do not eat slimy or pasty-textured pack-aged meats.

· Packages can be deceptive. Read the label carefully to see how many ounces of sausage or luncheon meat you are actually getting for your money. The total amount of meat inside the package is apt to fluctuate wildly from brand to brand, variety to variety, although the packages may appear similar in size.

· A good way to assure that your sliced luncheon meats or cold cuts are fresh is to purchase them fresh-sliced at the delicatessen counter of your supermarket. The thriftiest way possible to purchase cold cuts is by the slice, according to an estimate of how much you will need, rather than by the pound or fraction thereof.

· Unsliced sausage, particularly loaf liverwurst or luncheon meats, stays fresh in the refrigerator longer than sliced sausage.

· Combination packs of cold cuts and cheese generally cost more than the meat and cheese purchased separately.

· Freezing is not entirely recommended, especially for luncheon meats. If it is absolutely necessary, other varieties of sausage may be frozen, well wrapped, at 0° F. for three to six months. After that period of time, undesirable flavor alterations are likely to occur. It is always best to buy and use sausage as you need it.

· Choose a selection of at least three types of sausage and/or lunch meats for a cold meat platter. Let the senses of sight, taste, and smell be your guide. A variety of color, odor, texture, flavor, and shape will make your platter an attraction. And to make it spectacular, you may want to roll the meats, shape them into ribbons, or arrange them in concentric circles.

THE COLD CUT, LUNCH MEAT, AND SAUSAGE SELECTION

Alesandri is a garlicky, Italian member of the salami family. Its major ingredient is highly seasoned cured pork.

Alpino is another member of the salami group invented here in America.

Arles is a French version of salami. It contains coarsely chopped pork and beef seasoned with garlic.

Beef bologna contains only beef meat and has a pronounced garlicky flavor. Regular bologna contains both pork and beef.

Beerwurst is a cooked, smoked salami of German origin. It contains beef, pork, and garlic.

Berliner is a pork and beef sausage mildly flavored with salt and sugar.

Biroldo is a version of blood sausage.

Blood and tongue sausage contains cooked lamb and pork tongues and hog blood.

Bloodwurst or blood sausage is made of hog blood, pork meat, ham fat, gelatinous meats, salt, pepper, cloves, allspice, and onions.

* **Bockwurst** is a delicately flavored, highly perishable white sausage, consisting of fresh pork and veal, chopped chives, parsley, eggs, and milk.

Bologna is a mildly flavored, cooked, smoked, ground pork and beef sausage. It is ready-to-eat and second only to the frankfurter in popularity.

---

*Indicates that the product requires (or may require) thorough cooking.

* **Boterhamwurst** is Dutch-style veal and pork sausage.
* **Bratwurst** is links of pork, or pork and veal, distinctively flavored with herbs and seasonings. It can be bought both fresh and fully cooked.

Braunschweiger is a cooked, smoked liver sausage containing eggs and milk.

Calabrese is a dry, Italian, all-pork salami snapped up with red peppers.

Cervelat is a semi-dry sausage similar to salami. Its meat is more finely ground, however, and less highly seasoned than salami. Cervelat is sometimes called "summer sausage."

Chipped beef (see dried beef)

Chorizo is a Spanish dry link sausage made of highly spiced, hot tasting chopped pork.

Chub bologna is a smooth, emulsified mixture of beef, pork, and bacon.

Cold cuts is the term used to describe those prepared cold meats which are generally used for sandwiches and snacks.

Combination loaf consists of pork and beef with sweet pickles and peppers added.

* **Corned beef** is cured and spiced brisket of beef.

Cotto salami is cooked salami enhanced by whole peppercorns.

* **Country sausage** is a fresh pork sausage that is highly perishable. Sometimes beef or veal are also added to the smoked pork.

Deviled ham is a prepared specialty product made of finely ground ham and seasonings.

Dried beef is dehydrated beef—smoked, cured, and ultra-thinly sliced. It is sometimes referred to as "chipped" beef.

Dutch loaf is a cold cut comprised of pork and beef.

Easter nola is a mildly seasoned, salami-type, dry sausage of Italian origin.

Farmer cervelat consists of equal proportions of delicately seasoned beef and pork.

Frankfurters are made of cured, smoked, and cooked beef or pork, or a combination of the two. They are seasoned with white pepper, sugar, cloves, coriander, and nutmeg. The frankfurter is also known as a hot dog or wiener.

* **Fresh sausage** is a title covering the entire classification of uncooked sausages including the breakfast-type patty and link.

---

*Indicates that the product requires (or may require) thorough cooking.

Frizzes are a dry, highly seasoned sausage of coarsely chopped pork and beef.

Garlic sausage is similar in taste and texture to the frankfurter, but has a more pronounced garlic flavor.

Genoa salami is a pork sausage flavored with red wine and garlic.

German salami is more heavily smoked than the Italian-type salamis.

Goettinger is a hard cervelat of uniquely spiced pork and beef.

Goteborg is a heavily smoked, Swedish version of cervelat made of coarsely chopped pork and beef flavored with cardamom.

Gothaer is a German cervelat made of cured, finely chopped pork.

Ham and cheese loaf is a specialty product containing cubes of cheese encased in finely ground ham.

Ham bologna differs from regular bologna in that it contains diced, cured pork.

Head cheese is a cooked meat specialty product containing small pieces of pork head meat bound together by gelatin.

Holsteiner cervelat is a ring-shaped version of farmer cervelat.

Honey loaf is a mixture of pork, beef, honey, and spices. It can also contain pickles or pimentos, or both.

Hot dog (see frankfurter)

Hungarian salami is a mild version of salami.

Italian salami is a chewy, dry sausage of pork, flavored with red wine and garlic.

* **Italian sausage** contains cured, coarsely cut fresh pork and finely cut beef seasoned with fennel and wine. It may be purchased "sweet" or "hot" (seasoned with red peppers).

Kielbasa is a garlicky, highly seasoned Polish sausage of ground pork and beef.

Knackwurst is a fully cooked, garlicky variation of the frankfurter.

Knoblauch is another name for knackwurst.

Kosher salami is an all-beef salami flavored with garlic, mustard, coriander, and nutmeg.

Land jaeger is a black, wrinkly, heavily smoked Swiss member of the cervelat family.

Lebanon bologna is a coarsely chopped, heavily smoked beef bologna with a very distinctive flavor. It was first created in Lebanon—Pennsylvania.

* **Linguisa** is a Portuguese pork sausage cured in brine, seasoned with garlic, and spiced with cinnamon and cumin.

---

*Indicates that the product requires (or may require) thorough cooking.

Liver loaf is a sandwich-shaped liver sausage loaf, similar in flavor to liverwurst.

Liverwurst contains finely ground pork and liver combined with onions and seasonings. It is fully cooked and ready-to-eat.

Lola is a mildly flavored, dry Italian sausage seasoned with garlic.

Lolita is a mini-version of the lola link.

Longaniza is a dry Portuguese sausage with a flavor similar to chorizo.

Luncheon meat is the term used in reference to the various combinations of beef and/or pork and ham. Usually it is sold sliced and vacuum packed.

Lyons is a dry French pork sausage.

Macaroni and cheese loaf contains chunks of cheddar and pieces of macaroni in combination with ground beef and pork.

* **Mettwurst** is a German beef and pork sausage, smoked, cured, and spiced with coriander, ginger, and mustard. It is uncooked and has a smooth, spreadable consistency.

Milano salami is a dry, garlicky Italian salami.

Mortadella is an anise-seasoned, dry sausage containing pork, beef, and cubes of pork fat.

New England style sausage contains cooked, coarsely chopped, smoked pork.

Old-fashioned loaf is a cold-cut loaf consisting of pork and beef.

Olive loaf is a luncheon meat loaf containing whole, pimento-stuffed olives.

* **Pastrami** is the cured, smoked plate of beef. It is usually thinly sliced for sandwiches.

Peppered loaf contains pressed pork and beef flavored with cracked peppercorns.

Pepperoni is a dry, Italian-style sausage of beef and pork heavily seasoned with ground red pepper.

Pickle and pimento loaf is made of finely chopped pork and beef studded with pimentos and sweet pickles.

* **Pinkel** is a sausage made of beef, oats, and pork fat.

Polish sausage (see kielbasa)

Polony is an English version of bologna.

* **Pork sausage** is fresh, coarsely ground pork seasoned with sage, nutmeg, and pepper.

Salamette is the name for link salami.

Salami generally consists of coarsely ground pork, finely ground beef,

*Indicates that the product requires (or may require) thorough cooking.

and seasonings. Salamis vary in flavor and size according to their country of origin.

* **Salsiccia** is a fresh Italian sausage made of finely ground pork.

Scrapple is a specialty product of cooked, ground pork mixed with cornmeal flour.

Sicilian salami is yet another version of Italian salami originally created in Sicily.

Souse is pieces of pork meat in a vinegar-spiked gelatin base to which dill pickles, sweet red peppers, and bay leaves are sometimes added.

Smokies are smoked, cooked links of pork and beef spiced with pepper.

Straussburg is a liver and veal sausage containing pistachio nuts.

Summer sausage is a name that may be applied to all dry sausages and particularly to cervelat. Originally summer sausages were made in the winter to be eaten in the summer because of their ability to stand up without refrigeration. (We do not recommend trying it.)

* **Swiss sausage** is a fresh sausage similar to bockwurst.

* **Thuringer** is a pork sausage that may also contain some beef or veal. It can be purchased either fresh or cooked.

Tongue and blood loaf contains calves' tongues, pork fat, pork skins, and beef blood.

Vienna sausages are small "cocktail" type frankfurters.

* **Weiswurst** is a fresh, mildly spiced German sausage made of pork and veal.

Wiener (see frankfurter)

---

*Indicates that the product requires (or may require) thorough cooking.

Part VII

Consumer Power

Chapter 18

Selecting Self-Service Meat

*packaging, price, quality, less tender cuts…
supermarket service and sales*

THE INSIDE STORY ON PACKAGING: CRACKING CODES AND DEMYSTIFYING LABELS

From the consumer's point-of-view the perfect package for self-service fresh meat products has yet to be invented. The older styrofoam and fiberboard containers—overwrapped with transparent plastic film—enabled the prospective purchaser to see only that portion of the top and sides of the meat cut not covered up by the label. All too often one unwrapped the meat at home only to discover inordinate amounts of hidden waste.

Newer to supermarket meat cases, and a distinct improvement, are the clear plastic trays. Used in combination with transparent plastic film, they enable the customer to see both sides of the story. For the consumer these trays are not the last word on packaging, however. They have not yet eliminated the problem of bacteria build-up. Because they fail to absorb the juices of the meat, the moist environment inside the package provides a virtual breeding ground for bacteria. As the bacteria count rises, meat rapidly deteriorates.

Meat consumers also have to overcome the problem of the package label which is often used for reasons other than merely stating the name and weight and price of the product. While it is not the general policy of chain stores, we have found that the label is widely used by local outlets to conceal waste, fat, and bone. Until this practice is halted, the only way to be sure you are not paying for extra waste is to merely unwrap the package at the store.

Supermarket label

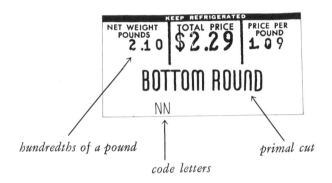

hundredths of a pound   primal cut

code letters

Illustrated above is a typical label indicating the cut, the primal cut from which it was derived, the price per pound, the net weight in pounds and hundredths of a pound, the letters that are the code for the day and the month that the meat was first offered for sale, and the total price of the purchase.

A complicated point that should be brought to consumer attention is the fact that on labels such as the one illustrated, any weight over a round pound is broken down into 100ths of a pound via the decimal system. One pound—as everyone knows—is 16 ounces. One tenth or .10 of a pound equals 1.6 ounces. Therefore, if the label indicates it contains 1.10 pounds, it does *not* contain 1 pound and 10 ounces, but 1 pound and $\frac{1}{10}$ of a pound, or one pound and 1.6 ounces. One and one half pounds would look like this on a label: 1.50 lbs. One and a quarter pounds like this: 1.25. Cracking this system will come in particularly handy when you are trying to figure the number of servings per pound. The average serving is usually somewhere between 4 to 8 ounces. In order to further help you decipher this decimal system, the following table is offered:

if the first decimal figure after the weight in pounds is:	=	the weight in ounces is:
.1		1.6 oz.
.2		3.2 oz.
.3		4.8 oz.
.4		6.4 oz.
.5		8 oz.
.6		9.6 oz.
.7		11.2 oz.
.8		12.8 oz.
.9		14.4 oz.
1.0		16 oz.

Keeping this chart in mind, one would be able to figure that 3.62 pounds as written on a label are actually three pounds and almost ten ounces.

Looking at the date code letters (or numbers, or a combination of both) on a ground beef package is another little trick that may come in handy if you want to find out if a cut other than the hamburger is really fresh. In most large stores hamburger is ground daily, sometimes twice a day to keep up with the demand. You can be sure, therefore, that the letters—which are used to conceal the date from everyone but the people who cut, wrap, and stock it—are up-to-date on the hamburger. A hamburger label almost invariably will bear the code of the day. If the code letters on the hamburger package correspond with those on the label of other meat purchases you are considering, the cuts will have most likely been put on the shelf that very day. If the code is different, however, chances are that the meat has been around awhile. Unfortunately this little code-cracking device cannot be extended to disclose precisely *how* long. We have it on good authority that the date codes are chosen in an unsystematic fashion. They are either deliberately picked at random or jumbled to make it impossible for anyone to decipher their message beyond that immediate day. The original and valid purpose of the code is to control the quality of the meat—not such a bad idea at all. Our only question is why can't these codes be designed to inform the consumer as well?

There are two advances in the packaging-labeling area that the consumer can look forward to in the near future. They are the label identification of ground beef according to fat content rather than primal cut, and nutritional labeling.

Under the new system for packaging ground beef, which at least one large chain store has already put into effect, hamburger would be labeled and priced according to the actual amount of lean meat it contains—the leanest being the most expensive. On the lowest end of the price spectrum, the ground beef could contain as much fat as the law allows (which varies from state to state), or the maximum amount acceptable: 30 per cent. The highest priced hamburger could contain as little as 15 per cent fat. Since much of the ground beef sold as "chuck," "round," or "top sirloin" contains scraps from other primal cuts and grinding virtually eliminates the differences in tenderness and often taste, this new labeling procedure represents a definite move in the direction of "truth in packaging"—and is wonderful for weight watchers too.

Nutritional labeling for meats is somewhat farther off in the future, more complicated, but not less important. Under the system, the per-serving amount of calories, protein, fat, carbohydrates, vitamins, calcium, and iron will be listed on a panel on the package. This new labeling concept will assist the consumer in making the proper choice between foods high and low in nutritive value, and enable one to make more intelligent buying decisions. Since nearly all meats are relatively high in nutritional value, bringing this point to the consumer's attention would seem to be advantageous from any point-of-view.

At the present, guidelines are being developed for the first category of goods that will be nutritionally labeled: frozen prepared dinners.

With all the technology we have going for us today, you would think that the purveyors of prepackaged, self-service meats could meet the final consumer challenge: a totally transparent label with both nutritional information and decoded dates.

GUIDELINES FOR VALUE: THE PRICE-PER-SERVING CONCEPT

In the supermarket, mistakes at the meat counter can be expensive. Yet the widely discussed and virtuous concept of unit pricing which allows one to buy foods on the basis of price per pound (per ounce, per pint, per quart) just is not effective when it comes to meat. The most important monetary consideration you can make at the meat counter is—the price per serving. It is your best guideline to value.

A serving may generally be figured on the average as 4 ounces of cooked lean meat. The actual number of servings a pound of meat will provide fluctuates widely from cut to cut according to the amount of fat, bone, and waste it contains. A pound of boneless meat will generally yield three or four servings, whereas a bony cut can provide as little as one portion per pound. A pound of lean hamburger, for instance, can be divided into four individual patties, but a pound of spareribs may only provide enough meat to serve a single hungry person—or 1½ people of moderate appetite. (Ground beef containing a good deal of fat will shrink and provide smaller, or fewer, individual servings.)

To figure out how much the meat costs per serving, divide the number of servings a cut yields per pound into the price per pound. Using this simple formula will help you to discover that the aforementioned lean hamburger at 89 cents per pound costs but 22 cents per portion, while the spareribs at the same price are far more expensive, costing 89 cents per portion for the hungry person or 59 cents for the smaller serving.

Another aspect of the price-per-serving concept to consider when judging the value of your meat purchase is this: the amount of shrinkage you should expect during the cooking process. The cooking method and the temperature at which the meat was cooked, as well as the degree of doneness, can all affect the actual number of ounces of cooked meat a pound of raw meat can deliver. As an example, without the bone, a pound of raw Porterhouse steak can actually provide as little as 7 ounces of cooked lean and fat meat, whereas a pound of raw round steak may yield up to 13 ounces of cooked lean and fat meat. The Porterhouse will, of course, be more tender with less cooking effort, but the round will feed more people. The question here is whether to indulge oneself and splurge or save and stretch the budget to cover other meals.

Traditionally people who trade with their personal local butcher have bought meat by the serving rather than by the pound. But the prepackaged, self-service supermarket meat section where the price per pound is clearly marked on every cut is here to stay. As such, it is the ideal training ground on which to test the price-per-serving theory and see if it doesn't cut your meat bills.

The following charts from the National Livestock and Meat Board offer a guide to the number of servings one can expect from the various cuts of meat and the cost per serving at various price levels.

Cost for a Serving of Meat At Various Price Levels

COST PER POUND	APPROXIMATE COST PER SERVING							
	1½ Servings per Pound	2 Servings per Pound	2½ Servings per Pound	3 Servings per Pound	3½ Servings per Pound	4 Servings per Pound	5 Servings per Pound	6 Servings per Pound
$ .39	$ .26	$ .20	$ .16	$ .13	$ .11	$ .10	$ .08	$ .07
.49	.33	.25	.20	.16	.14	.12	.10	.08
.59	.39	.30	.24	.20	.17	.15	.12	.10
.69	.46	.35	.28	.23	.20	.17	.14	.12
.79	.53	.40	.32	.26	.23	.20	.16	.13
.89	.59	.45	.36	.30	.25	.22	.18	.15
.99	.66	.50	.40	.33	.28	.25	.20	.17
1.09	.73	.55	.44	.36	.31	.27	.22	.18
1.19	.79	.60	.48	.40	.34	.30	.24	.20
1.29	.86	.65	.52	.43	.37	.32	.26	.22
1.39	.93	.70	.56	.46	.40	.35	.28	.23
1.49	.99	.75	.60	.50	.43	.37	.30	.25
1.59	1.06	.80	.64	.53	.45	.40	.32	.27
1.69	1.13	.85	.68	.56	.48	.42	.34	.28
1.79	1.19	.90	.72	.60	.51	.45	.36	.30
1.89	1.26	.95	.76	.63	.54	.47	.38	.32
1.99	1.33	1.00	.80	.66	.57	.50	.40	.33
2.09	1.39	1.05	.84	.70	.60	.52	.42	.35
2.19	1.46	1.10	.88	.73	.63	.55	.44	.37
2.29	1.53	1.15	.92	.76	.65	.57	.46	.38
2.39	1.59	1.20	.96	.80	.68	.60	.48	.40
2.49	1.66	1.25	1.00	.83	.71	.62	.50	.42
2.59	1.73	1.30	1.04	.86	.74	.65	.52	.43
2.69	1.79	1.35	1.08	.90	.77	.67	.54	.45

The servings per pound are only a guide to the average amount to buy to provide 3 to 3½ ounces of cooked lean meat. The cooking method and cooking temperature, the degree of doneness, the difference in the size of bone in the bone-in cuts and amount of fat trim are some of the factors that vary and will affect the yield of cooked lean meat.

BEEF

Steaks

Cut	Servings per pound
Chuck (Arm or Blade)	2
Club	2
"Cubed"	4
Filet Mignon	3
Flank	3
Porterhouse	2
Rib	3
Rib Eye (Delmonico)	3
Round	3
Sirloin	2½
T-Bone	2
Top Loin	3

Roasts

Cut	Servings per pound
Rib, Standing	2
Rib Eye (Delmonico)	3
Rump, Rolled	3
Sirloin Tip	3

Pot-Roasts

Cut	Servings per pound
Arm (Chuck)	2
Blade (Chuck)	2
Chuck, Boneless	2½
English (Boston) Cut	2½

Other Cuts

Cut	Servings per pound
Brisket	3
Cubes, Beef	4
Loaf, Beef	4
Patties, Beef	4
Short Ribs	2

Variety Meats

Cut	Servings per pound
Brains	5
Heart	5
Kidney	5
Liver	4
Sweetbreads	5
Tongue	5

PORK

Chops and Steaks

Cut	Servings per pound
Blade Chops or Steaks	3
Boneless Chops	4
Fresh Ham (Leg) Steaks	4
Loin Chops	4
Rib Chops	4
Smoked (Rib or Loin) Chops	4
Smoked Ham (Center Slice) Steaks	5

Roasts

Cut	Servings per pound
Ham (Leg), Fresh, Bone-in	3
Ham (Leg), Fresh, Boneless	3½
Ham, Smoked, Bone-in	3½
Ham, Smoked, Boneless	5
Ham, Smoked, Canned	5
Boston Shoulder (Rolled) Boneless	3
Loin, Blade	2
Loin (Rolled), Boneless	3½
Loin, Center	2½
Loin, Smoked	3
Picnic Shoulder (Bone-in) Fresh or Smoked	2
Sirloin	2
Smoked Shoulder Roll (Butt)	3

Other Cuts

Cut	Servings per pound
Back Ribs	1½
Bacon (Regular), Sliced	6
Canadian-Style Bacon	5
Country-Style Back Ribs	1½
Cubes (Fresh or Smoked)	4
Hocks (Fresh or Smoked)	1½
Pork Sausage	4
Spareribs	1½
Tenderloin (Whole)	4
Tenderloin (Fillets)	4

Variety Meats

Cut	Servings per pound
Brains	5
Heart	5
Kidney	5
Liver	4

LAMB

Chops and Steaks

Cut	Servings per pound
Leg Chops (Steaks)	4
Loin Chops	3
Rib Chops	3
Shoulder Chops	3
Sirloin Chops	3

Roasts

Cut	Servings per pound
Leg (Bone-in)	3
Leg (Boneless)	4
Shoulder (Bone-in)	2½
Shoulder (Boneless)	3

Other Cuts

Cut	Servings per pound
Breast	2
Breast (Riblets)	2
Cubes, Lamb	4
Shanks	2

Variety Meats

Cut	Servings per pound
Heart	5
Kidney	5

From *Teaching About Meat*, National Live Stock and Meat Board, 1972.

CONVENIENCE VERSUS QUALITY (AND COST)

It has been emphasized in this book that convenience foods are expensive, although undeniably they can come in very handy if one isn't very hungry and doesn't have time to prepare a meal.

The U.S. Department of Agriculture federal standards require that frozen dinners include a minimum of 25 per cent meat. Yet according to studies, a prepared dinner from a store's freezer section can cost you more than twice as much as the same exact meal (in the same proportions) prepared from scratch at home. Under these same federal standards, a dish such as prepared veal parmigiana is required to contain only 40 per cent breaded meat product in sauce—or at least 28 per cent cooked meat (without the breading or the sauce). In such a case, it seems to us that you are likely to end up with "Meat so dressed and sauced and seasoned that you didn't know whether it was beef or mutton—flesh, fowl or good red herring."*

In this era of high meat prices, even those expensive cuts of steaks and chops can be regarded unofficially as convenience foods in some respects. They are basically so simple to prepare but oh so high in price per portion.

If time is of the essence and one's food budget extremely tight, and yet there are hungry faces to feed, far better to reach into your freezer for those frozen-but-honest hamburger patties you prepared when you bought ground meat. And if time allows, but your budget still won't, try experimenting with some of the less-than-tender cuts by applying some of the tenderizing techniques.

MAKING THE MOST OF THE LESS-THAN-TENDER CUTS

So you are in the supermarket, hovering over the meat section, wavering on the decision between the luscious-looking loin and the less costly shoulder chops. What should you do?** Remember, any cut from a part of the animal that moves (is mobile) is likely to contain fibrous connec-

*George DuMaurier, *Trilby*.
** Read Chapter 13, "Lamb—The Retail Cuts."

tive tissue and cost less. The less costly cuts are *no less nutritious* than their higher-priced counterparts, although they may require longer and a slightly more complicated cooking method to make them tender enough to eat. Here are some of those less-than-tender cuts:

BEEF:	chuck, brisket, flank, foreshank, rump, neck, plate, short ribs
FRESH PORK: '	back ribs, end-cut chops, fresh shoulder, butt, shoulder, shoulder chops
SMOKED PORK:	boneless shoulder butt, ham shank, ham hocks, picnic ham
LAMB:	breast, neck, shank, shoulder roast, shoulder chops
VEAL:	breast, shoulder roast, shoulder chops
VARIETY MEATS:	heart, kidney

None of the cuts above, in fact no cut of meat, need be too tough to eat if it is properly cooked. Proper cooking is the simplest and the best of all the tenderizing techniques. Any cut that can be roasted can also be pot roasted. Any cut that can be broiled can be pan broiled or fried. The reverse, however, is not necessarily true. And although it certainly wouldn't be recommended that you put a filet mignon to the test, *any cut at all* will take to some sort of moist cooking: pot roasting, braising, simmering, or stewing. It is into this latter category that many of the less-than-tender cuts fall. Even the undergraded traditionally tender cuts of less than USDA Choice quality often take better to moist cooking methods than to roasting or broiling. If you have qualms about the quality of a cut (is it mobile? is it Choice grade?), consider a recipe that calls for one of the tenderizing moist-cooking methods. If, however, you still wish to roast or broil, make sure the meat is at room temperature before you start to cook it to reduce the shrinkage. Sear meats to be broiled close to the flame. Then complete the cooking as far from the direct heat as possible. Roast meats at a lower than average oven temperature.

Another technique of tenderizing meat for roasting which is especially effective with lean cuts of beef, pork and veal, is the nearly lost art of larding. Once you get the knack of it, larding takes very little time, and makes up for the lack of marbling in the meat. Undergraded meats or those cuts with little or no exterior or interior fat will especially benefit by the extra juiciness and flavor imparted by the inserted lardoons. "Lardoons" is the name for the slivers of pork fat. These scraps of pork fat are available by request at most meat counters. "Larding" is the name of the technique by which the lardoons are inserted into the meat by means of an 8- to 10-inch long hollow needle. This needle is essential to the success of larding—and unfortunately may be somewhat hard to come by except at certain restaurant supply stores and gourmet cookware departments.

In order to lard, the fat must first be cut into lardoons ⅛ inch wide and up to 2 inches in length. Each strip is threaded, one at a time, through the end of the neeedle with a tiny bit of the tip of the lardoon popping out the top of the needle. The needle is then pulled through the meat at one inch intervals, releasing the lardoons into the interior of the meat. When the meat is roasted the lardoons melt and make the meat juicy.

An even easier way to keep a roast from drying out is to simply secure extra fat to the exterior of the roast before popping it into the oven. If bacon or pork fat is used, the method is called "barding" and barding does about the same thing for meat that basting does. Bacon will impart some flavor to the meat, however, which can even be a distinct advantage in the case of many meat loaves. You may simply wish to rub the roast with a little olive or cooking oil prior to cooking as an alternative.

Marination is yet another satisfactory tenderizing technique. The meat is immersed for 2 to 24 hours or more in a subtly flavored marinade. For short periods of time marination is best accomplished at room temperature—for longer periods under refrigeration.

A marinade is an acidic liquid rather like French dressing, composed of either lemon juice, vinegar, or wine usually in combination with olive oil and other flavoring ingredients such as garlic, herbs, and spices. The enzyme action in a marinade helps to break down the tough fibrous tissue of the less-than-tender cuts and secondarily to lend a unique and interesting flavor to the meat. Arabs and Greeks have been employing the tenderizing effects of lemon on lamb (and tough goat meat!) with success for centuries. You may want to consider doing the same with lamb kabobs cut from the shoulder rather than the leg. Tough cuts of beef can also be deliciously tamed by the subtle flavor and enzymatic action of wine or wine vinegar.

Many recipes for marinades exist. It is recommended that the home-made variety be used rather than the commercially prepared papaya-based products which tend to break down the cell walls of the meat and make the texture pulpy rather than truly tender.

Pounding meat can also be tenderizing to some extent, particularly with cuts such as veal scallops or chuck minute steaks. You may have noticed a butcher pounding a cut with the side of his knife to help flatten it, make it of equal thickness throughout, and help break down the fibrous tissue. You can do likewise with utensils specifically designed for the purpose—or even the edge of a saucer. One should be cautious not to tear the meat, whatever method is employed.

THE SECRET SUPERMARKET
MEAT SERVICE—
WHERE AGGRESSIVENESS PAYS OFF

In times past when one shopped at the small, neighborhood butcher, verbal communication was the key to purchasing a cut. Today, the marketing of meat—like the selling of so many things—is a mass-production concept. But if you are accustomed to shopping for meat at a self-service supermarket, and the great majority of us are today, somewhere behind that meat section—albeit shrouded in secrecy—there is very likely to be a meatcutter hidden, a person who is available only by knocking or ringing a bell. Many of these meatcutters are protected from actual customer contact by sliding windows, but they can and should be contacted. In reality they are there to serve you. Getting to know them—or better yet, impressing them with your expertise—is bound to pay off in lower meat bill dividends and more meal variety too.

One highly knowledgeable, enterprising, and aggressive shopper we know started something of a minor rebellion at her local chain store. When the packaging prevented her from really seeing the spareribs she was considering purchasing, she simply removed the outer wrapping and then the spareribs from the package. Dissatisfied with the trim on the ribs, she took them to the meat counter and asked for service. She asked the cutter to remove the excess fat, citing the fact that she would end up throwing it away. Not only did the cutter oblige her request, but he reweighed the newly trimmed spareribs and, before repackaging, repriced them according to the new, trimmed weight. Our shrewd shopper ended up paying less for a better buy. Observing the lady in action, her astonished fellow shoppers decided to follow her example. Before long the supermarket chain store meatcutter was trimming ribs for a small crowd. While this delicious example is extreme, and widespread similar actions could probably cause reverse rebellion among the ranks of the meatcutters fraternity, it does, we think, offer a good idea of the kind of effective service you can expect and get at self-service meat counters if you are armed with the proper knowledge and nerve. A more realistic approach to this available cutting service would be a request to subdivide a large, cumbersome cut in order to provide several meals for your family. For this service, it may be the policy of the store to request a nominal charge. It is worth a nominal charge. Cutting cumbersome cuts can provide a "theme with variations" (and you'll have the bonus of fewer leftovers with which to cope).

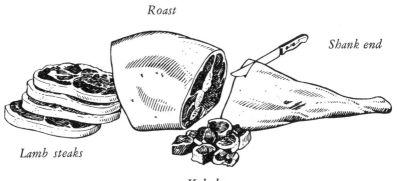

Roast

Shank end

Lamb steaks

Kabobs
American Meat Institute

A leg of lamb, for example, offers several interesting possibilities— especially if the weight is high (8 to 10 pounds) and the price per pound low. To obtain three meals for a family of four, simply ask the meatcutter to slice four lamb steaks from the sirloin end of the leg. Ask him also to remove the lower part of the shank end and cut the meat into kabobs. (You can, in fact, cut the kabobs yourself at home.) Finally, for meal #3, the center portion of the leg is reserved for roasting.

A family of four can also feast on a whole shank half of smoked ham—three times. Simply ask the meatcutter to remove a generous

Shank

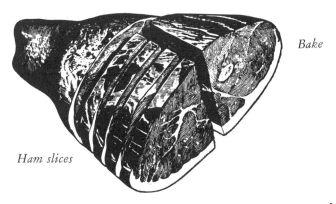

Bake

Ham slices

American Meat Institute

shank portion for meal #1. That section will be delicious and hearty simmered with cabbage and potatoes. You and a sharp knife can easily do the rest. Just divide the remaining part of the ham into a large section for baking and four generous ham slices for yet another meal.

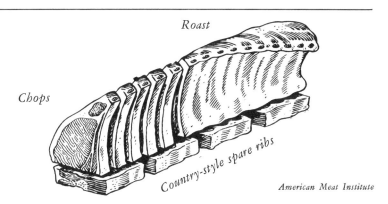

Roast

Chops

Country-style spare ribs

American Meat Institute

The rib end of a loin of pork roast can present the cutter with something a little more challenging, and present you with three more fresh-cooked dinners of four servings each. Ask the cutter to saw through the backbone section, leaving an inch of meat on the bones. Have him divide the backbone into individual portions—and you'll have country-style spareribs. Meal #2 is pork chops (sliced on request) with the remainder of the rib end portion providing a pork roast.

Similar possibilities of extra meals are also available with beef rib roasts and Porterhouse steaks. We suggest, however, because they represent a major purchase, that you buy these cuts on sale. (See page 000.)

MEAT SALES:
TO BUY . . . OR NOT TO BUY?

Studies have shown that supermarket specials and sales are indeed valid. They can offer the diligent shopper an average savings of 20 per cent and even up to 35 per cent—over an extended period. Since meat is the item on which the greatest percentage of the food budget is expended, it just makes sense (and can save dollars and cents) to take advantage of them if they are legitimate. There can be certain pitfalls, however.

Sales at supermarkets are deliberately planned events. They are intentionally designed to draw heavy traffic into the store. Local newspaper supermarket advertisements beg for business by offering truly reduced prices and some of the more consumer-oriented and evolved chain stores have begun to issue advance bulletins on upcoming sales events. This seems to us to be a smart practice from the points-of-view of everyone concerned. These advance bulletins allow the smart consumer to plan his menus around the upcoming specials and to curb his impulse buying which, of course, all stores encourage. For the retailer these bulletins are an invitation to his customer to come into the store and buy (sometimes dangling the bait of cheaper cuts of meat under fancy labels). The informed customer is on to the fact, however, that a "his 'n her" steak could be just about any cut of beef that will serve two, and that "butter steak" is cut from the chuck.* (Many states are beginning to outlaw this advertising practice and requiring that the primal cut be named.) The alert shopper is also aware of exactly what he wants before he enters the store—and exactly how it should look.

All shoppers should regard meat sales as an invitation to look at the goods, but not necessarily to buy them. They should try to envision and compare the quality of the cut featured on sale with that same particular cut at its regular price. Sometimes the sale pork chops are end cuts rather than center cut, and rib roasts might not be of the more select first cuts. An alert shopper can detect the difference.

*See glossary on cuts of beef in Appendix 3.

The trim may tend to be quite different also, so that the sale cut could contain more wasteful fat and bone than with its normal trim. This could cut sharply into the overall savings. Also roasts prepared for sale will not always be "oven ready." That means that the little extra cutting step which facilitates the difference between a roast easily carved and one that is impossible to carve has been omitted during the sale. Sloppy carving at the table can only contribute once again to the amount of waste. The situation can easily be remedied, however. For the cutter it is usually only a matter of cracking the chine bone. Even if the roast is a sale item, the customer has the right to ask for this service before purchasing it—either without charge or for a very nominal fee.

A good consumer practice is also to look for the secondary or "sleeper" special. It is the responsibility of the chain store meat manager to see that all of his meat moves evenly from the shelves and into the hands of the customer. For that reason, when an item such as beef top round roasts are the featured special, the anatomically adjacent rump may also be priced lower per pound. Both are cut from the back haunch of a steer, which is how the primal cut is usually received by the market before it is subdivided. This holds true for all cuts obtained from a single primal cut of the animal anatomy, but particularly of beef chuck and beef round—yet another reason to bone up on your meat-related anatomy.

Another sale practice left to the discretion of the local supermarket meat manager by the chain store executives is just how far he will "merchandise" a particular cut before he features it on sale. Sales on rib roasts and beef loins are particularly enticing and draw eager customers into the store. Once there, however, the disappointed shopper may find that all the cuts are heavy—the rib roasts weigh 12 to 14 pounds or more, the whole loins up to 14 pounds. There are savings to be had by purchasing these giant cuts, but in most cases it means a substantial initial capital outlay which the average consumer cannot always afford. The cuts are also usually just too large for his immediate needs. So, the consumer on a tight budget will end up purchasing something else. And the store will still benefit in two ways by the large sale cuts left unsold. They have served their original purpose of drawing customers into the store and the cutter need only recut them into more compact and acceptable sizes and sell them at a higher price per pound after the sale is over.

For the average family, if freezer space and budget will allow, sales on properly trimmed rib roasts—regardless of size—are definitely a time to buy. All one needs to take advantage of the sale is the ready cash and the expertise to ask the chain store's in-house meat cutter to divide the large cut into several smaller ones which will provide several meals. For this service, however, expect a small nominal charge.

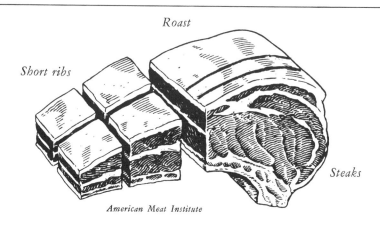

Roast

Short ribs

Steaks

American Meat Institute

A 12-pound rib roast, as an example, will provide about two servings per pound, and can be subdivided into one company-sized roast, several good-sized rib steaks, and short ribs for braising—three entirely different meals.

When Porterhouse steaks—even the poorly trimmed type with extra long tails—are on sale at a substantial reduction it would be advantageous for a family to buy three or more. Each Porterhouse can be dissected at home into the makings for three different meals. The filet is removed for one meal, the strip loin becomes another, and the tail can either be cut into cubes for beef stew or barbecued and eaten as is. (To accomplish this yourself at home, see Chapter 20, "The Step-by-Step Guide to Cutting Meat.") In this case even the bones can be used as the basis of a good beef stock.

Another sort of instant, last-minute sale designed to move merchandise quickly may occur at the end of a day or a weekend. The cuts featured are likely to be of the smaller, more expensive variety, which have outlived their shelf life—nuisance merchandise for the retailer. Rather than try to trim or grind them, he would rather reduce their price and move them quickly off the shelves. While not at the peak of their freshness, these cuts can be quite acceptable if eaten the day they are purchased.

Chapter 19

❦❦❦❦❦

Buying Beef in Bulk— and the Chilling Truth About Freezer Plans

A re there substantial savings to be had by buying frozen beef in bulk? The answer to that question is yes—and no. In most cases the savings on specific cuts will just about equal, or be only slightly less than, those savings you get by taking advantage of a good supermarket special. As detailed in Chapter 18, "Selecting Self-Service Meats," an alert supermarket shopper with ready cash in hand can make truly substantial savings by capitalizing on specials. Some of the best buys are to be had on the larger bulky cuts, purchased with the intent of freezing them for later use.

However, unless those special bulky cuts are subdivided into smaller ones prior to home freezing, it is impractical to freeze and store them in the ordinary home freezer for any length of time. The average home freezer simply does not maintain temperatures low enough to completely freeze the interior of a really thick, large cut. Its quality will, therefore, rapidly diminish.

In this sense, legitimate concerns which specialize in selling large bulk orders of frozen meats at near wholesale prices have an advantage over the shopper left to her own devices and her home freezer. The advantage is called "blast freezing." Blast freezing is a process which exposes meats to sub-freezing temperatures far, far lower than the ordinary home freezer can achieve. Blast-frozen meats are frozen quickly at a very low temperature—usually −40° below zero Farenheit—under pressure. Because the bacterial growth on the meat is thus arrested so rapidly, the meat will keep fresh longer than meats bought fresh for freezing at home. Consequently, buying frozen meat in bulk from a legitimate specialist does have certain fringe benefits.

In recent years, however, some of the most blatant examples of consumer fraud and deception involved frozen "family food plans" and the sale of a freezer. The deal was this: you got the freezer—free—if you purchased the frozen meat and other foods on a long-term monthly payment plan. The hitch was that the quality of the meat and food was inferior and not to the tastes of the customers, so the freezer ended up being overpriced, and often repossessed if the purchaser was unable to meet the monthly payment. Unscrupulous companies who offer such deals specialize, of course, in preying upon the poor. Thankfully in this era of consumer awareness, such concerns are beginning to disappear.

Another alluring deal that has been offered to consumers in the past involved the sale of a side of beef at a per pound price ridiculously low —or seemingly so. What was not made clear to the prospective customer was that the side he purchased was all but "on the hoof." By the time waste, fat, and bone was eliminated and the side divided into usable cuts, the actual price per pound had escalated due to the diminishing returns.

"Bait-and-switch" advertising is yet another gimmick designed to deceive prospective customers. Certain frozen meat concerns have been known in the past to "bait" customers and bring them into the store with advertised prices so incredibly low he cannot possibly resist. Often these "deals" did not even exist but were simply created for the purposes of the advertisement. Once the unwitting customer is in the clutches of the unscrupulous dealer, he "switches" him over with high-pressure tactics to another much higher-priced frozen meat package.

If you are entertaining the notion of making a big investment in beef, it is best you know everything there is to know about beef beforehand. One of the chief factors to bear in mind before you buy in bulk is the fact that you must either own a home freezer or have access to a freeze locker. You will also want to consider that you may end up with a lot more of the lesser cuts than you really want. If you purchase a 300-pound, untrimmed side of beef, for instance, you can count on about 82 pounds of unusable waste. Of the remaining 218 pounds, only slightly more than 19½ pounds will be rib roasts and steaks, 37 pounds cuts from the loin (Porterhouse, T-bones, sirloins). The round will provide some good roasts, but more than 160 pounds of the side will be cuts from the chuck, plate, and flank as well as the round that maybe you didn't really want to begin with.

Before making any major beef purchase, check out as best you can the reputation of the dealer from whom you plan to purchase it, to make certain you can rely upon him. Then sit down with him and go over all the details so that you will know exactly what you are getting. Make sure that the price he quotes per pound includes the cost of processing (cutting, wrapping, freezing)—or at least find out how much extra you will have to pay—it can be as much as 8 cents to 10 cents per pound—and figure that cost in the total price. Make sure the primal cuts are subdivided into your family's favorite cuts. If your preference is hamburger rather than stew meat, now is the time to say so.

Find out what the yield grade is of the section of beef you plan to buy. "Yield grade" is the U.S. Department of Agriculture measure of the actual amount of usable meat a carcass will provide—particularly the preferred retail cuts. The USDA Yield Grades are 1, 2, 3, 4, and 5. They are based to a large extent on the amount of exterior fat around the

lean. Yield grade 5 would offer the least amount of usable meat per carcass—only 65.9 per cent or less. Yield grade 1 would result in 79.8 per cent or more of retail cuts. Most meat is designated at yield grade 3 or about 72 per cent.

The yield grade is unimportant when purchasing individual retail cuts because it does not reflect quality. A USDA Choice rib roast could come from a carcass which was any one of the five yield grades. However, when buying in bulk it can make a substantial difference in the actual amount you get for your money.

If you are limiting your selection to a quarter of beef, also keep this fact in mind: A hindquarter will give you more roasts and steaks, but with a forequarter you will get more lean meat at a lower cost.

Above all, stay away from any contractual, long-term agreements or frozen food plans. Read the advertisements but steer clear of anything that smacks of deception or commercial bribery. Become an expert on . . .

READING THE MISLEADING

Advertisements are valuable. They can inform and instruct. But they can also lead a shopper into the trap of expecting a great deal more than he is actually getting. This is particularly true of some "specials" offered by large-scale local frozen meat outlets. The best advice that one can give a prospective frozen meat customer is to read the small type in an advertisement as well as the large type—and even then remain dubious.

Typically confusing is an advertisement which recently ran in a New York daily newspaper for a multi-branch frozen meat outlet's "summer sale." The ad contained a chart similar to the one illustrated below. The chart is representative of a side of beef. The code letters are used to indicate the areas of the various primal cuts:

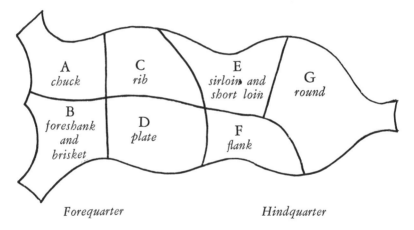

The following explanation was printed in very small type underneath the chart:

A. Chuck Steak, Pot Roast, Chuck Tender. B. Shank Cross Cuts, Stew Meat, Ground Beef. C. Standing Rib Roast, Rib Steak, Club Steak. D. Short Ribs, Plate Beef, Rolled Plate. E. Sirloin Steaks, Porterhouse, T-Bone, Sirloin Tip. F. Ground Beef, Flank. G. Top Round Steak, Eye of Round, Rolled Rump, Bottom Round.

A most intriguing section of this advertisement was printed in tiny type near the bottom of the page. It plainly stated that all the meat orders were sold at "hanging weight" and were subject to trim loss in "cutting and wrapping." "Hanging weight" means that the primal cuts are totally uncut and untrimmed—or in their original state fresh from the slaughter house. It also means that the consumer would eventually end up paying for pounds of hidden waste, fat, and bone that he'll never even see.

At the top of the page in bolder type face, the dealer sought to seduce the consumer away from this point by offering as a come-on the choice between 20 pounds of chickens [sic], 10 pounds of hot dogs, pork loin, or bacon for only $1 with the purchase of 100 or more pounds of beef. Considering the pounds of waste you will be paying for, the "bonus" chicken you choose is no bargain.

The same advertisement offered in bold type the following kinds of "specials." They are, we think, typical of the kind of misleading deals one should try to avoid. They are the sorts of high-pressure gimmicks designed specifically to get you to buy more beef than you actually need or would ordinarily eat.

USDA CHOICE
LOIN & RIB SECTION
69¢ per lb.
C, E, D, and F on chart
average weight: 145 to 175 lbs.

If one were, in fact, receiving only the loin and rib sections, 69 cents per pound would be a fantastically low price. What the large type doesn't tell you is that the wasteful flank and plate sections are also part of the deal. (This fact becomes clear if you bother to decipher the code in small type at the bottom of the ad by correlating it with the chart of the side of beef.) After all, the purveyor has to make his profit somewhere. The flank and plate, which are mostly soup, stew cuts, and waste, are sold at a relatively low price everywhere. Combined, they would account for more than one-third of the meat in this package deal. This kind of bargain is definitely to be avoided.

Another offering in the same ad was this:

USDA CHOICE
STEAK & ROAST ORDERS
99¢ per lb.
for one full side of beef plus all the
Sirloin, T-bone, Porterhouse, Rib Roast,
Delmonico, Short Ribs and Plate from
other side. (A side plus C, E, D, and F on
chart)

Even if you had the space to store all the cuts from a side of beef plus C, E, D, and F on the chart (decoded read: rib, loin, flank, and plate)—we'll bet this "steak and roast order" would provide more stewing cuts and hamburger than you ever imagined you had bargained for. In fact, discounting the waste and trim, what you actually end up getting is only about 50 per cent roasts and steaks; the rest is soup meat, stewing cuts, and hamburger. You will also note that the average weight of this particular package has been omitted. Depending upon the size of the steer (the average side is about 300 pounds), you had better be prepared to plunk down the price of at least 400 pounds of beef and probably more.

Another USDA Choice "small freezer special" allowed the consumer to have for only . . .

49¢ per lb.

letters A and B on the chart
average weight: 45 to 70 lbs.

. . . or shank meat, brisket and chuck—much of which is hamburger, stewing cuts and waste. (Aside from the fact that even without the shank included, an entire USDA Choice chuck usually weighs more than 70 pounds.) If your freezer is small it seems simply silly to stock it with nothing but secondary cuts of meat.

If you are intent on stocking a freezer it is suggested that you steer clear of gimmicks and "bonuses" or anything "free" and seek out a reputable dealer (there are more than 7,500 of them all over the country) who can supply truly fine beef at fair prices. One caution: in some areas this concern may be difficult to locate. Its owners may not bother with much advertising—because they find they can do a brisk business by relying on their reputations.

Chapter 20

✣✣✣✣✣

The Step-by-Step Guide to Cutting Meat

a dozen do-it-yourself, cost-cutting ideas you can accomplish in your own kitchen

If you own one thin-bladed, well-sharpened, five-inch knife, you can cut your own meat. It is *that* easy! Although it will not be necessary for most of the demonstrations described herein, some of them will require the use of a slender, 10-inch knife as well. Thus equipped you are ready to begin to be your own butcher.

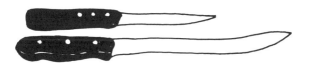

There are many bonuses in learning some simple meat cutting methods. Boning up on cutting techniques can stretch the meat you serve, add variety to your menus, and make your meals more glamorous and creative. But best of all, cutting meat at home can save you money. Generally the savings will be quite substantial. And, once you get the knack you can experiment with cuts other than the dozen demonstrated here.

Caution: before taking knife in hand, it might be best to read these valuable tips for the uninitiated:

1. Raw meat cuts most easily when it is cool. At room temperature it tends to get mushy and unmanageable. Therefore, before you attempt to cut it, it is best to chill the meat in the freezing compartment of your refrigerator for about ten minutes. When the meat is removed, it should be about as firm as a ripe tomato, and far easier to handle. If you are aiming for thin, even slices, this rule holds especially true.

2. For easy handling, cut chilled meat on a butcher's block, cutting board, or plain wooden surface. It will prevent the meat from being slippery and help you to have a steady hand.

3. Relax. Grip the knife in either hand in the position most comfortable to you. Hold the meat firmly with the other hand. Always cut away from yourself and pull the meat towards you. If you are inexperienced, work slowly.

4. After cutting meat, wipe it with damp paper towels and dry it. Then, either use it immediately, wrap it loosely to store in the refrigerator, or wrap it securely for the freezer. (See Chapter 21, "Storing and Freezing Meats.")

All of the meats used in the cutting demonstrations that follow can be readily purchased or ordered in advance at most supermarkets. Just follow the simple steps outlined below. You'll be surprised how really easy it is. (The demonstrations are illustrated for right-handed knife-wielders. If you are left-handed, you will have to make the proper adjustments.)

DEMONSTRATION #1:
THE PORTERHOUSE EXTENSION

You purchase: two 2½- to 3-pound Porterhouse steaks, 1 to 1½ inches thick

You cut: filet mignon, two 5- to 8-ounce portions
shell steaks, four 5- to 6-ounce portions
cubed beef, 1 to 1½ pounds

Follow the same cutting procedure for both steaks:

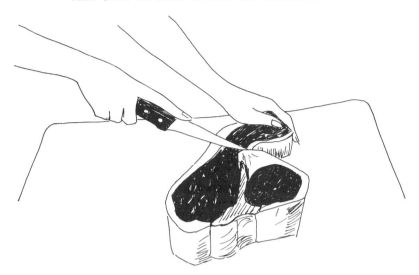

Step 1.   Place each steak flat on the cutting surface. Following the diagram, insert the tip of your knife in the narrowest part of the steak and slice straight across to remove the tip or tail. One smart stroke of the knife will do it.

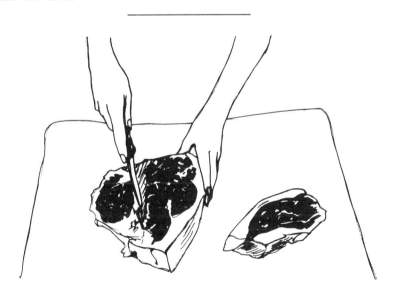

Step 2.   To remove the filet mignon (otherwise known as the tenderloin or eye), cut in as closely to the top part of the T-bone as possible, working slowly with the tip of your knife.

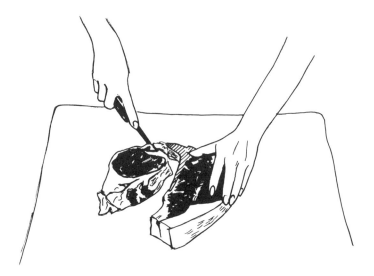

Step 3. Continue to scoop along and around the bone to about its center. At this point you will be able to remove the tenderloin.

Step 4. Trim it of excess fat. Each tenderloin will weigh from 5 to 8 ounces. The section remaining with the bone intact is known as a T-bone or club steak.

Step 5. Using the same cutting method you used to remove the tenderloin, cut in as close to the bone as possible, and remove it entirely. The result will be a 10- to 12-ounce boneless shell or strip steak. Trim of excess fat.

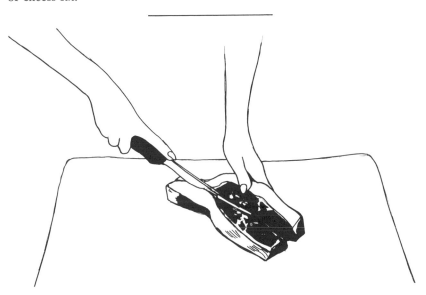

Step 6. Turn meat over and around, and slice vertically from the thick end. Cut each shell in half if you wish to provide two 5- to 6-ounce supper-sized servings.

Step 7. Cut the tails that were removed first into 1-inch cubes for stew or kabobs. There will be enough cubed beef to serve four people.

DEMONSTRATION #2:
THE 3-RIB ROAST TRANSFORMED

You purchase: the first three ribs of a semi-boned rib roast. (A self-service supermarket generally sells rib roast semi-boned. This means that the chine bone, back bone, and short ribs have been removed. To be certain of purchasing the first three ribs, see Chapter 4, "The Beef Roast from Coast-to-Coast.")

You cut: rib steaks, six 12-ounce steaks, or eight
9-ounce steaks, or twelve 6-ounce steaks

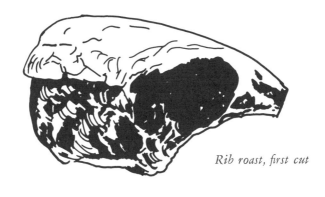

Rib roast, first cut

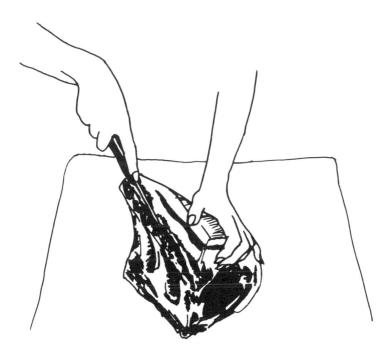

Step 1. Following the diagram, with the tip of your knife cut
across the three exposed ribs and down to their thick end.

Step 2. If this maneuver is successfully accomplished, all three bones can easily be removed in one piece.

Step 3. Place the now-boneless roast fat side up and shave some of the fat off the top with the blade of your knife. Leave just enough fat to provide flavor and moisture. A thin layer should do the trick.

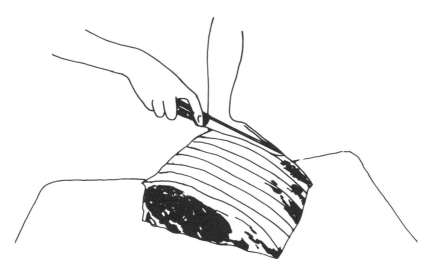

Step 4.   After removing the excess fat, leave the roast upright and divide it into eight steaks this way: score seven long, vertical, equi-distant lines on the fat with the tip of your knife as cutting guides.

Step 5.   Letting the marks be your guide, proceed to slice straight down through the roast vertically. A long, slender knife will make a cleaner cut and it is easiest to start from the thick end. If you prefer thinner steaks, divide the roast in twelvths. For thicker steaks, divide in sixths.

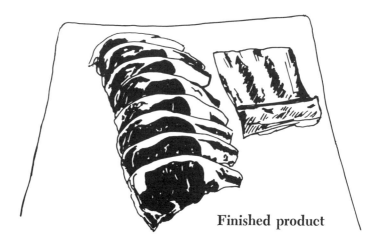

Finished product

Step 6. Remove any excess exterior fat.

Note: Save the rib bones you have cut away. They make a great addition to the stock pot.

DEMONSTRATION #3: THE 3-MEAL RIB ROAST

You purchase: rib roast, one end cut, 10 to 12 pounds, including the last 2 or 3 ribs with the chine bone and the back bone removed. (Have the short ribs sawed off by the meat cutter and reserve them.)

You cut: rib roast, one 6 to 8 pounds, top rib pot roast, one 2½ to 4 pounds, short ribs, 2 or 3

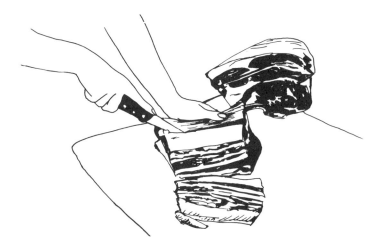

Step 1. Separate the short ribs from one another by cutting between each rib bone.

Step 2. With your knife, trim away any excess fat.

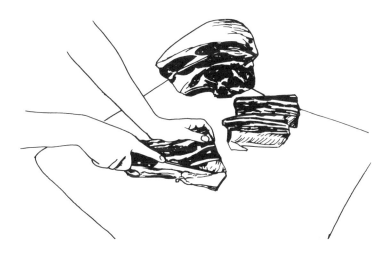

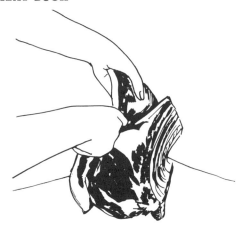

Step 3. Separate the pot roast meat from the whole rib roast. Stand the roast up as in the diagram. Insert the tip of your knife into the extreme end of the roast where the short ribs were removed, just above the large section of fat which divides the two layers of meat.

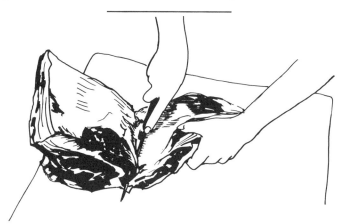

Step 4. As in the diagram, continue to cut over and around the fatty section, lifting the upper layer from the lower until the entire upper layer can be completely removed.

Step 5. Trim the excess fat from the meat you have removed, and starting at one end, roll the meat into a cylindrical shape.

Step 6. To help the top rib pot roast hold its shape, encircle it with kitchen twine and tie it several times.

Step 7. The remaining piece of meat, including the rib bones, needs no tailoring. Any excess scraps of fat remaining may be added to the exterior surface of the meat to keep it juicy and flavorful as it roasts.

Finished product

DEMONSTRATION #4: CHATEAUBRIAND AND FILET MIGNON—AT A SAVINGS

You purchase: a whole, well-marbled 7- to 10-pound beef tenderloin

You cut: Chateaubriand, (one 5-pound roast), and 1 pound stew meat, or one 3-pound roast, 1 pound stew meat, and filet mignon, four 8-ounce servings

---

Whole beef tenderloin

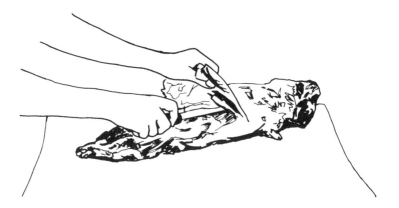

Step 1. Following the diagram, place the slender, elongated whole tenderloin fat side up. There is no fat on the bottom. Tuck the tip of your knife under the fat at the thinnest end of the tenderloin. Work toward the thick end and remove all of the top layer of fat by using short, even strokes of the knife.

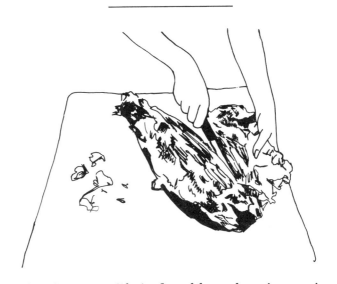

Step 2. As you cut, lift the flap of fat, and continue cutting toward the thick end of the tenderloin. It does not matter if the fat layer breaks or crumbles as it is removed, but you should try to avoid cutting into the meat of the tenderloin if possible. Cut fat off along the length of tenderloin and continue to remove all exterior fat. Save the fat you have removed. (See note.)

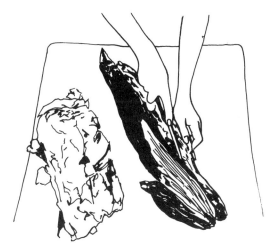

Step 3. When the tenderloin is free from fat, cut away the heavily membraned side piece as in the diagram. Use this meat for cutting into cubes for stew or kabobs or grinding.

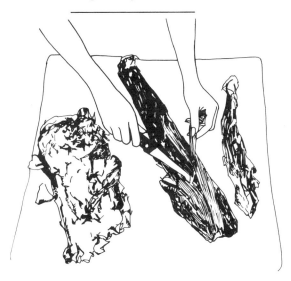

Step 4. A thin, irridescent membrane will now be visible along the thick end of the tenderloin. This must be removed if the beef is to be tender and easy to carve. Tuck your knife underneath the membrane at the center of the tenderloin and work toward the thick end. Using short strokes, cut away from yourself and lift the membrane until all of it is removed.

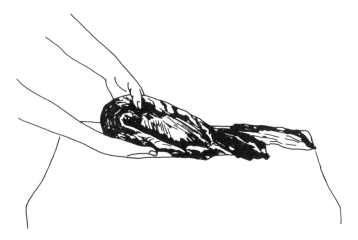

Step 5.   Turn the roast over so that the leaner side is on the bottom. To shape the roast evenly fold the narrowest end of the meat underneath the lean side of the tenderloin.

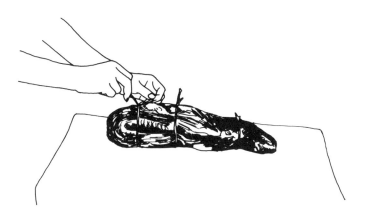

Step 6.   Secure the flap to the roast by tying it with a loop of kitchen string. Also tie the roast in the center and at its thick end for neat, shapely results. (If you plan to encase the semi-cooked roast in pastry, remove the string before you do.) The Chateaubriand is now ready to cook. It will be about 4 inches wide, 10 to 12 inches long, and provide enough meat to feed eight people generously.

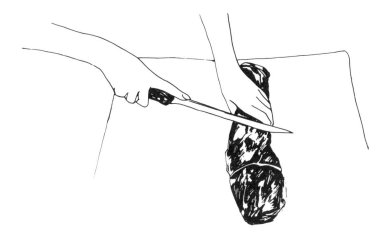

Step 7. If you prefer a smaller roast and individual-portion filet mignons, simply slice the tenderloin vertically down the center. Save the thick end for roasting and tie as you would a larger roast.

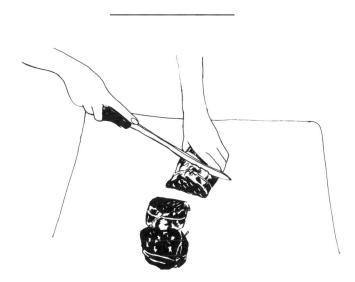

Step 8. Tie the tail end with a loop of kitchen string in four places before cutting into individual portions in order to keep each filet mignon secure and neat. Then slice the thin end vertically into four individual-portion filet mignons.

Finished product

Note: If a meat roast lacks its own surface fat and has scanty marbling, an extra layer of fat is often tied to its exterior in order to keep it moist while it cooks which helps to make it tender. This extra layer of fat is called "barding." Although pork fat and blanched bacon are often used for barding, the excess fat removed from the tenderloin can also be used. Here's how: place the pieces of fat—crumbles, slivers, and all—between two pieces of waxed paper. Place a bottle, pan, or other weighty object over the paper and press down. A rolling pin also works well for this. As you press down and iron over the fat, it will flatten, and become uniformly thick. To use, simply cut it into the width and shape desired, and secure it to the exterior surface of a roast with a piece of kitchen string. Barding fat may be stored in the refrigerator for up to a week, or in the freezer for six to eight weeks.

DEMONSTRATION #5: MAKING THE MOST OF CHUCK ROAST

You purchase: a boneless, 4-pound chuck roast (first cut) (this cut is sometimes called an "inside chuck roll")

You cut: roasts, one 1½-pound, well-marbled
                chuck eye roast, and one lean pot roast,
                or four 6-ounce steaks, and up to ten
                sandwich-sized steaks

Boneless chuck roast

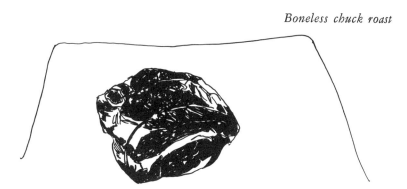

Step 1. Chill the chuck in the freezer for ten minutes to make it firm. Remove it from the freezer, and place the roast on a flat surface. If it is tied, remove the string. Turn the roast over and look for the little gap in the solid meat. This is where the rib bone was removed.

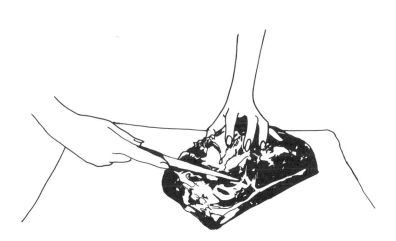

Step 2. As in the diagram, start at the gap with the tip of your knife, and cut straight across following the natural dividing line of fat. This will divide the roast into two pieces.

Step 3. One piece will be lean, the other highly marbled. The lean piece may be left whole, marinated, and pot roasted. It will provide enough meat to feed six or more people. The marbled piece of chuck may be broiled or roasted as is and will feed four people.

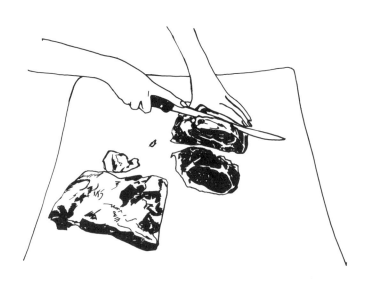

Step 4. As an alternative, the marbled piece may be divided into four individual steaks. (See diagram.)

Step 5.   The lean part of the chuck may be divided in half.

Step 6.   Slicing down vertically, each half may be cut into five sandwich-sized steaks.

Finished product

DEMONSTRATION #6:
STRETCHING SIRLOIN STEAK

You purchase: two 2- to 3-pound center-cut sirloin steaks
with filet, each 1 to 1½ inches thick

You cut: filet mignon, four small portions sandwich
steaks, 1 pound, eight to twelve steaks
cubed beef, 1¼ pounds for kabobs

Follow the same cutting procedure for both steaks:

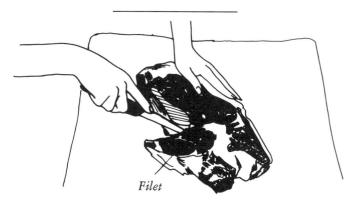

Filet

Step 1.   Place the steak flat on the cutting surface and insert the tip of your knife into the meat at the end of the flat bone.

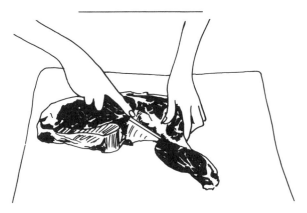

Step 2.   Following the diagram, turn meat and cut in to remove the filet (or eye). Trim any excess fat from the filet and rechill it in the freezer for about ten minutes.

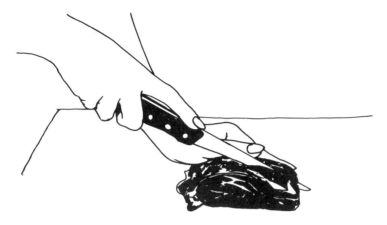

Step 3.   After the meat has chilled, split each filet through the center horizontally with your knife. This will make four small filet mignon portions. Flatten the four filets with the palm of your hand to make them uniformly thick throughout. To store, wrap each filet individually.

Step 4.   To remove the balance of the meat from the bone, push the knife straight along the edge of the bone and peel the bone away from the meat.

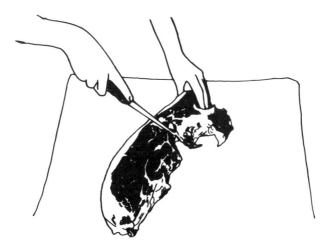

Step 5.   Remove the tail by slicing along the natural seam separating it from the sirloin's center. Reserve the tail for cubing.

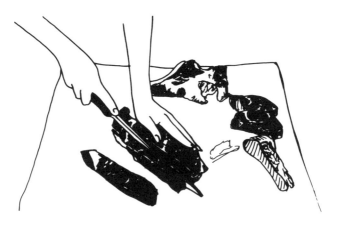

Step 6.   Use the meat from the center of the sirloin for sandwich steaks. Trim the center portion of excess fat, then, following the diagram, cut horizontally into thin slices, beginning at the oval end of the meat. There will be 4 to 6 of these sandwich-sized portions in a sirloin steak.

Step 7. Trim any excess fat from the tail, and cut it into 1-inch cubes. The tails from both of the sirloins will provide enough beef cubes for kabobs or stew to make from four to six servings.

Finished product

DEMONSTRATION #7: A LAMB CHOP TREAT FROM LAMB STEW MEAT

You purchase: lamb breasts, two 1½- to 3-pound including the flank with the thick-end bones cracked, or two 1- to 2½-pound without the flank, and 1½ pounds of ground lamb

You cut: lamb chops, eight double thick, stuffed

Step 1. At the store, have the flank portion of the lamb breasts removed and ground. If the store is self-service, the bones at the thick end of the breast will generally have been cracked. If not, have them cracked.

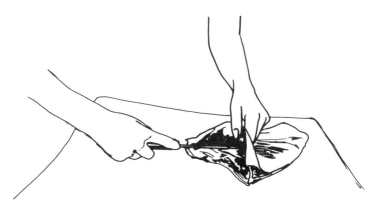

Step 2. Following the diagram, place the breasts bone side down. Tuck the tip of your knife underneath the thin layer of meat at the widest end of each breast. Probe underneath the layer of meat with your knife, working the full length and width of each breast until you have inserted pockets about 10 inches long and 4 inches wide. Some people find a long, thin knife makes this easier.

Step 3. If you wish, season the ground lamb with herbs, garlic, and spices. Divide the meat in half and insert one half into each of the breast pockets.

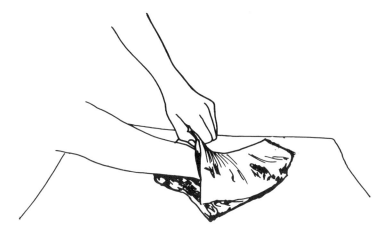

Step 4. Push the ground meat all the way down and spread it evenly inside the breast. As the pockets are stuffed, the breasts of lamb will expand.

Step 5. Turn both breasts of lamb bone side up. According to how they were originally cut, each will contain from ten to thirteen rib bones. To cut the double-thick chops, start at the exposed end of the third rib (see diagram), and slice the length of the bone all the way through the breast. Continue to cut between every second or third rib. Cut each breast of lamb into four equally thick chops. The chops may be roasted, broiled, or braised.

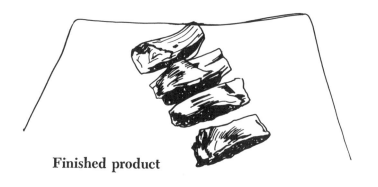

Finished product

DEMONSTRATION #8:
RACK OF LAMB REVISED

You purchase: rack of lamb, one whole

Note: In supermarkets this cut is almost always cut and sold as chops, so you may have to order it in advance. By purchasing the rack, you should be able to save 10 to 20 per cent. Have the meatcutter divide the rack into two parts and remove the backbone.

You cut: Roast, one Frenched rack, chops, eight Frenched lamb chops or four double-thick rib chops

Half rack of lamb

Step 1.   Place one half of the rack fat side up. Score the fat deeply until the tip of your knife touches the rib bone, working straight across the width of the rack, at a distance of about 2 inches from the top of the rib bone.

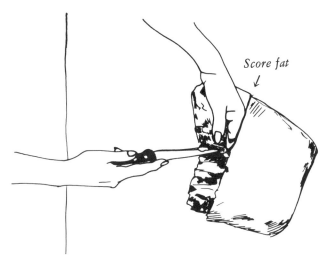

Score fat

Step 2. Continue to cut horizontally with your knife. With your other hand, lift the entire strip of fat from the tip of the bones. This is purely waste fat. Discard it. There will now be 8 or 9 rib bones exposed, depending upon how the rack was cut originally.

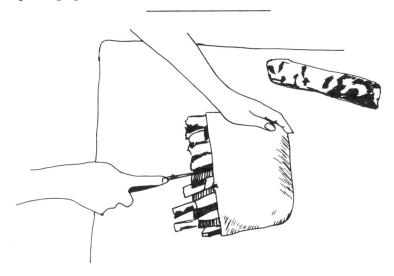

Step 3. With your knife, pare away all the fat and meat in between each rib bone. Make sure each protruding rib is virtually bare of meat for about 1½ to 2 inches from its tip. This is called "Frenching" a rib bone.

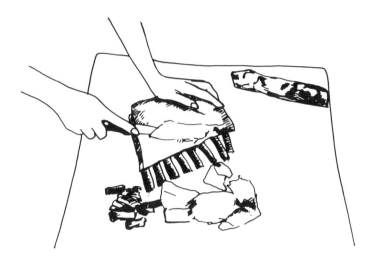

Step 4. Trim excess fat from the roast.

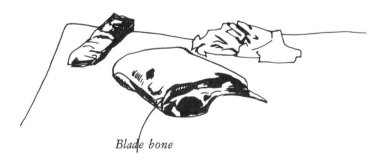

Blade bone

Step 5. Cut away the tiny blade bone indicated in the diagram. The roast is now ready to pop in the oven. (See note.) The roast will feed four.

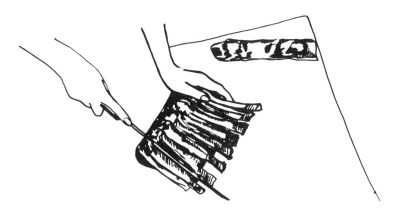

Step 6.   Place the other half rack of lamb on its flat surface. To make Frenched lamb chops, proceed as above, steps 1 through 5. Turn the roast fat side down and slice between each rib vertically to divide the half rack into 8 or 9 chops. (It is not necessary to "French" chops. Frenching serves no purpose other than making a regular rib chop a bit fancier. But if you like the little touch it adds, you can save money by doing it yourself, rather than having it done by a butcher.)

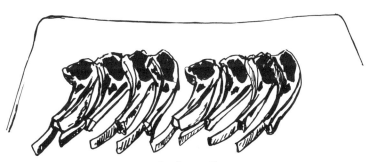

Finished product

Note: To keep rib bones from charring while the lamb cooks, cap each with a piece of aluminum foil. If desired, present the roast at the table with a paper frill atop each rib.

Note: For double-thick rib chops, omit steps 1, 2, and 3. Remove excess fat and blade bone, as in steps 4 and 5. Then place the second half rack on its flat surface, and slice vertically between every other rib. You will get four fat chops.

DEMONSTRATION #9:
FOUR-WAY PORK

Pork rib roast

You purchase: pork rib roast, one center cut 8 to 10
pounds with the back bone removed

You cut: roast, one Frenched loin, or chops, ten to
thirteen Frenched, or six to eight butterfly chops or eight cutlets.

If you wish the Frenched pork roast, follow steps 1 through 4 from
the preceding demonstration: *rack of lamb revised.* You can cook this
roast as is, but if you prefer Frenched pork chops, leave roast fat side
down and with your knife, slice vertically between each rib, allowing one
rib per chop, (see illustration page 229, step 6.) Allow 1 to 2 chops
per serving depending on size. For butterfly chops, proceed as follows.

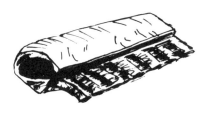

Step 1.   Place the whole roast fat side up. Score the fat and
remove it.

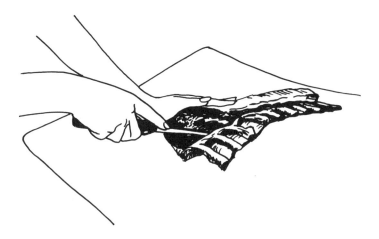

Step 2. Tuck the tip of your knife into the meat at the point the excess fat has been removed.

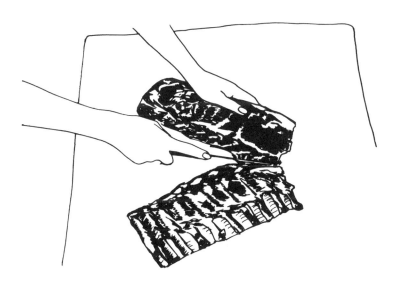

Step 3. Cut in close to the bone along the entire length of the roast, and remove the center core of meat from all the bones in one large piece. This cut is known as the shell of pork loin.

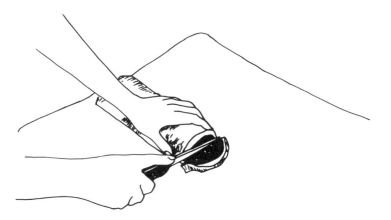

Step 4.   Place the shell fat side up and trim away all of the excess fat. Starting ½ inch from the end of the shell, slice down vertically and about three quarters of the way through.

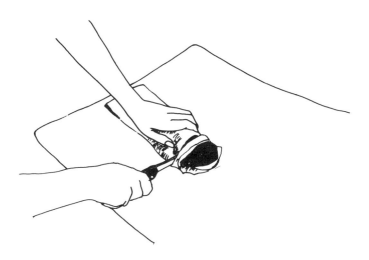

Step 5.   Move your knife another ½ inch to the left—this time about one inch from the end of the shell—and slice vertically all the way through. The result is a butterfly chop.

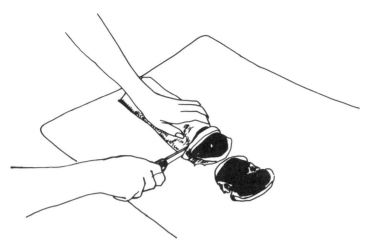

Step 6. Continue to slice the shell vertically at ½-inch intervals, alternating the three-quarter slice with the full slice. The entire shell of pork loin will make about eight butterfly chops.

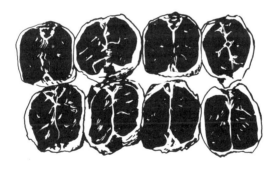

Butterfly chops

Finished product

To make pork cutlets, spread the sides of the unstuffed butterfly pork chops on a flat surface. Pound the cutlet with the palm of your hand to make it flat and evenly thick. You may want to flour or bread the pork cutlet before cooking.

DEMONSTRATION #10:
DOUBLE VEAL MEAL

You purchase: breasts of veal, one 7 to 9 pounds with a
pocket, one 5 to 6 pounds (have thick-
end bones cracked on both)

You cut: a roast, breast of veal stuffed with veal—
enough to feed 12 people

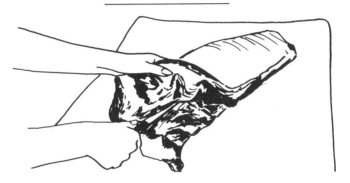

Step 1.   Place the smaller breast of veal bone side down. Starting
at either end (or even in the middle) cut and scrape the meat away from
the bone with the blade of your knife. Use sharp strokes and cut away
from you, lifting the meat as you cut. Neatness does not count here. The
meat you cut from the bone will be concealed in the pocket of the larger
breast of veal, so cut away with confidence. Reserve the bones you cut
away. (See note.)

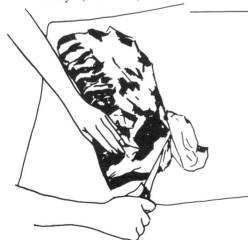

Step 2.   Trim the excess fat
from the boneless veal.

Step 3. Using your hands, shape the meat into a cylinder that will fit neatly into the pocket of the larger breast of veal. Press any cutting scraps remaining into the cylinder also.

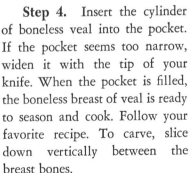

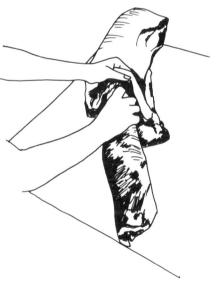

Step 4. Insert the cylinder of boneless veal into the pocket. If the pocket seems too narrow, widen it with the tip of your knife. When the pocket is filled, the boneless breast of veal is ready to season and cook. Follow your favorite recipe. To carve, slice down vertically between the breast bones.

Note: These bones are nice to nibble if they still contain a little meat. Divide the bones by slicing between each rib. To cook, scatter the rib bones around your breast of veal in the roasting pan about midway through the roasting time.

DEMONSTRATION #11:
HAM THREE WAYS

You purchase: shank end of ham, one 6 to 8 pounds
(have butcher remove the hock)

You cut: six ham steaks, diced ham, 1 pound soup
bone, one large ham hock, 1–1½ pounds

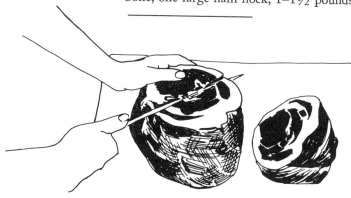

Step 1.   Place the ham down on its largest cut surface. One side of the ham contains bone, the other side is boneless. Following the diagram, insert your knife as close to the bone as possible.

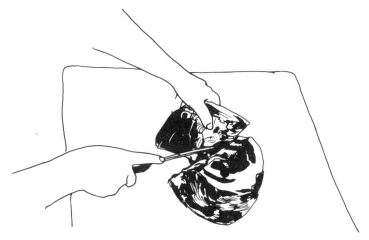

Step 2.   Slice down and as close to the bone as you can, and remove the bony side of the ham.

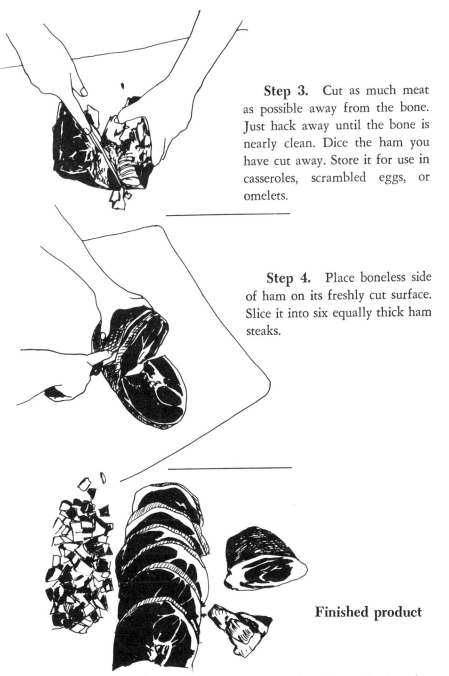

Step 3. Cut as much meat as possible away from the bone. Just hack away until the bone is nearly clean. Dice the ham you have cut away. Store it for use in casseroles, scrambled eggs, or omelets.

Step 4. Place boneless side of ham on its freshly cut surface. Slice it into six equally thick ham steaks.

Finished product

Step 5. What you have left is a ham hock and a nearly clean ham bone. How about brewing up a pot of pea soup?

DEMONSTRATION #12:
SMOKED HOCK "CHOPS"

You purchase: ham hocks, four smoked of similar size, 1 to 1½ pounds each (Make sure they are from the hindshank end and not the foreshank, which looks the same but is smaller and referred to as knuckles.)

You cut: three hock chops, diced ham, ½ to ¾ pound, smoked ham skin (for flavoring)

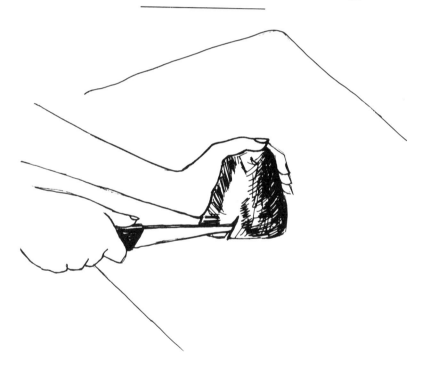

Step 1.   Remove the hard-smoked outer skin from the ham hocks. Following the diagram, grasp the shank end of a hock firmly in one hand and puncture the skin at the wider end of the hock with the tip of your knife.

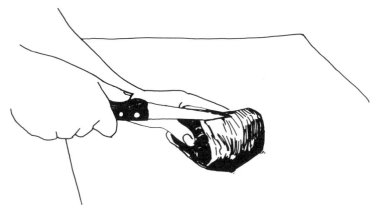

Step 2. Insert the knife, blade side up, just inside the puncture. Using sharp, upward cutting strokes, continue to split the skin in one continuous line until you have freed it completely from the bone at the opposite shank end.

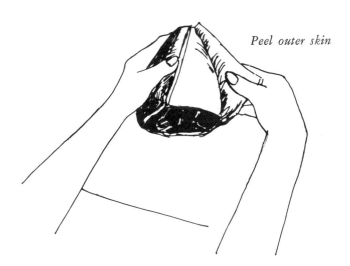

Peel outer skin

Step 3. Continue to grasp the shank end in one hand, and with the other hand peel the hard outer skin back in the one piece until the interior meat is fully exposed and the skin is completely removed. Reserve the skin (see note). Repeat steps 1, 2, and 3 until all the hocks have been skinned. These skinless hocks, which we call hock chops, will be much more tempting and easier to eat than ham hocks that are unskinned.

Step 4. Select one of the hocks for de-boning. Insert the tip of your knife into the hock until it touches the bone. Cut around the bone, freeing the meat, until all the meat has been removed from the bone. Neatness does not count.

Step 5. Dice the hock meat finely and use it in omelets, scrambled eggs, soups, or stews.

Note: Diced, smoked ham hock skin is a tasty flavor booster for hearty soups, greens, and beans. Smoked ham hocks are fully cooked and need only be reheated by any cooking method or in combination with other foods. More good news: the average hock will cost less and provide more lean meat than an entire pound of bacon.

Chapter 21

Storing and Freezing Meat

safekeeping for the various cuts and kinds...
freezer organization... the hazards of freezing
...wrapping, cooking, and thawing methods...
how long meat lasts

All meat is perishable. Ideally it should be eaten as soon as possible after it is purchased. Most of us, however, have neither the inclination nor the time to shop for it everyday. Much of our shopping must be done in advance. Meat storage, therefore, is important.

If meat must be saved—where in the refrigerator, in what, and for how long? If meat is to be at its peak when prepared for the table, other safekeeping aspects to consider are its proper care and preparation for storage, when and if meat can (or should) be frozen, and if your equipment is up to it.

For quick reference, a chart is provided at the end of this chapter listing conservative recommended maximum storage periods for particular meat. The storage and freezing of meat can be tricky, however, because the same rules do not necessarily apply to all cuts and kinds. For example, the storage *methods* differ among these five basic types of meat:

1. Fresh
2. Cooked
3. Cured, smoked, and ready-to-eat
4. Canned
5. Frozen

1. **Fresh meat** includes all the cuts of beef, pork, veal, lamb, and ground meat as well as variety meats and fresh sausage. All fresh meat should be stored promptly in the refrigerator. An airy, uncluttered refrigerator that allows for the free flow of air will keep all foods fresh longer. The temperature inside of most refrigerators hovers around 38° F. Some refrigerators have a special compartment or container especially designed to keep fresh meats. In absence of this container, it is best to store meats in the coldest section of the refrigerator without freezing them. Self-defrosting refrigerators are particularly hard on meat. Their principle is to remove the moisture from the air inside the refrigerator—and they also tend to remove the moisture from meat. If you own a refrigerator of this type, you should consider using the meats sooner than would be necessary with an ordinary refrigerator.

If meat is purchased prepackaged, as most meats are in a self-service supermarket, it may be left in its original wrapper if you plan to eat it within a day or two. However, if the meat is wrapped in ordinary market (or butcher's) paper, the paper should be loosened to allow some air circulation to reach the meat. A certain amount of air drying of the meat's surface is desirable during storage. A dry surface inhibits the growth of bacteria. Some of the more progressive butchers are using oxygen impregnated paper these days. It actually releases oxygen to meat and helps it to last a little longer. Ask your butcher if he uses this type of paper. If so, it will be safe to keep his meat snugly wrapped.

Remember, a moist meat surface encourages bacteria to multiply. That is why fresh meat should never be washed before it is stored. You may, if you feel it is necessary, wipe meat with damp paper towels just before you cook it—but always dry it, too.

2. **Cooked meat** can include leftovers, prepared-in-advance dishes, meat stocks, gravies, sauces, soups, and stews. All cooked meat should be cooled rapidly before it is stored. Simply place the cooked meat, preferably on a rack, in the airiest spot available and wait for it to cool. If liquid, stir it occasionally, or divide it into smaller portions to speed up the cooling process. The quickest way to accelerate the cooling process is to place the cooked meat in a big bowl or pot large enough to accommodate it. Then, partially submerge the pot in cold water. In any case,

the cooked meat should be at room temperature and covered in airtight wrapping material before it is placed in the refrigerator. Meat dishes warmer than room temperature will cause the temperature inside your refrigerator to rise—and hasten the spoilage of other foods. Inside the refrigerator, cooked meats not properly cooled will also form condensation inside their wrappings. This moisture will drop down on the meat and make it soggy, as well as speed the growth of bacteria.

Large cuts of meat, such as roasts, will dry out less rapidly if they are left intact. If storage space is at a premium, however, the large bones may be removed. Some prepared-in-advance casseroles will actually improve in flavor quality if they are allowed to sit well covered for a day in the refrigerator. Generally all cooked meats should be eaten within the week they are prepared.

Stocks, stews, soups, and gravies will keep best in screw-top glass jars or airtight plastic containers. Some experts claim that if properly stored, well-concentrated stock will stay well in the refrigerator for up to two weeks. In order to keep it, a fairly complicated procedure which requires re-boiling to further concentrate the stock, then re-cooling and re-storing it about every three days is suggested. Since stock is easy to freeze and thaw, if freezing space is available, we think it far simpler to freeze it.

3. **Cured, smoked, and ready-to-eat meats** include hams, bacon, pastrami, corned beef, cold cuts, smoked pork, and sausage. In the refrigerator, these products will generally outlast fresh meats. They keep up to a week, or slightly more in some cases. This is due to their extra processing. Cured, smoked, and ready-to-eat meats should always be kept in tightly covered or airtight containers to prevent them from drying out and their odors from escaping and penetrating other foods. Meat that is vacuum packed or purchased in a plastic film wrapper will keep best in the original wrappings. Sliced bacon will not stay truly fresh for more than three or four days after the package has been opened. Slab bacon will last considerably longer. A cold cut specialist in our neighborhood tells us that ready-to-eat meats will be tastiest if they are removed from the refrigerator and left to stand at room temperature for an hour before they are served. This would, of course, not hold true for frankfurters or other ready-to-eat sausages which are usually heated before they are eaten.

4. **Canned meat** falls into two categories: those that require refrigeration and those that do not. Most of the larger canned hams will, many of the smaller ones will not. Most canned bacon doesn't, but some canned luncheon meats do. The best rule is to check the label on the can, and carefully follow its instructions. Canned meats do have a fairly long shelf life whether they require refrigeration or not. It is not ever advisable to freeze canned meats.

---

5. **Frozen meat**—The home freezing of meats is not complicated although it does require a certain amount of preparation, equipment, and expertise if it is to be successfully accomplished. Meat can only be safely frozen for a long period of time at temperatures of 0° F. or below. The temperature in most refrigerator ice cube compartments is considerably higher. Therefore, the ice cube section cannot double as a freezer and should not be used as such. Unless you own a freezer, or a refrigerator with a separate outside door opening on a totally separate freezing unit, the procedure should not be attempted. You may, however, keep fresh meat in an ice cube compartment if you follow the general rule for freezing and eat it within a week.

Frozen meat ideally should be used before the expiration of the time recommended for its storage. Be sure to consult the chart at the end of this chapter for the amount of time it is advisable to freeze the various cuts. Beyond that time a rapid deterioration in quality will undoubtedly occur.

Refreezing of thawed meat is not recommended and should be avoided, although it is not harmful in and of itself. Refrozen meat will simply suffer a loss of quality and much of its juices. Partially thawed meat, however, with ice crystals still clinging to the surface, can usually be successfully refrozen without a vast discernible difference in quality.

Above all, freezing offers economical convenience. There is something reassuring about a well-stocked freezer. It allows one to cook and buy meat in quantity and to save it for future use. It also permits one to take advantage of seasons and the supermarket sales. By following the simple rules detailed below, freezing will help to preserve the nutritive value, texture, flavor, color, and aroma of most fresh meats better than any other storage method.

Preparing meat for the freezer—For peak efficiency, the meat you plan to freeze should be ready to cook before it is put in the freezer.

According to your estimate of how it will be served, meats should be divided into individual, family and company-sized portions. Add spice to ground meats and form them into ready-to-cook patties or meat loaves—but never add salt before freezing. Salt is omitted because it inhibits the speed of the freezing process itself.

Planned-in-advance casseroles and main course meat dishes are best if cooked about 15 minutes short of their required total time before they are cooled and frozen. This allows for the amount of time required to reheat them without overcooking. Stews, hashes, meatballs, curries, and the like will freeze best if generously sauced. Cooked meat roasts freeze best in the largest pieces possible (without jeopardizing your freezer space situation).

Wrapping, packaging, and labeling—Prepackaged meat in a self-service sort of wrapping may be frozen as is for a week or two. Be sure to check and see that there are no punctures in the package, however. If so, or if the meat must be stored for longer periods, it will need to be rewrapped or overwrapped.

To ensure that meats are properly packaged, a moisture-vapor proof material that is sturdy enough to withstand handling at sub-freezing temperatures must be used. Ordinary waxed paper or butcher's market paper simply will not do. Heavy-duty plastic bags, aluminum foil, heavy-duty polyethelene film (such as Saran wrap), or specially coated freezer paper work best for the bulky cuts. Many manufacturers produce freezer-to-oven pots and pans which are perfect for planned-in-advance dishes. Air tight plastic, snap-lid coffee cans, or aluminum foil containers work well too. Save those packages from other foods you purchase and recycle them in your own freezer. Another handy item that will help keep packages snug and airtight is low-temperature tape.

The classic freezer wrap is illustrated on the next page. To execute it, simply place the meat in the center of the sheet of freezer paper—shiny side up if using the coated kind. Gather the opposite sides together and turn them down about three times in one-inch folds until the wrap is snug against the meat. Iron across the package with the palm of your hand to force out the air. Then fold the ends together and over as if you were wrapping a gift. Secure the ends with tape.

If using heavy-duty plastic bags for freezing, place the meat in the

Freezer wrap

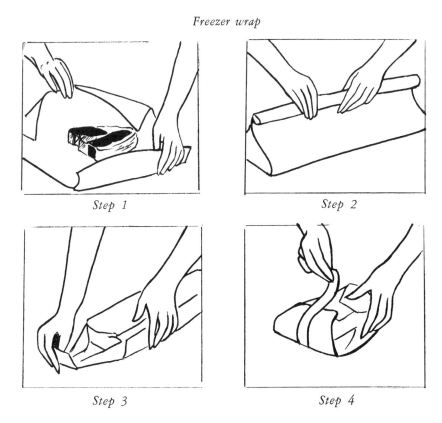

Step 1   Step 2

Step 3   Step 4

bag, gather the excess plastic together semi-loosely with your hand and hold the open end to your mouth—then inhale. Quickly twist the bag and tie off the top of it. By inhaling the air, the bag deflates and becomes airtight and the plastic clings to the surface of the meat.

Heavy-duty aluminum foil (or polyethelene wrap) requires no tape or tie. Be sure to cut a sheet large enough to cover the piece of meat generously. Place the meat in the center of the sheet, gather the sides together, and make a lock seam, but be careful not to gather the foil up too tightly. Press the excess foil down and mold it to the meat to exclude the air. Fold the ends up tight to the meat and avoid punctures. The perfect freezer wrap is both airtight and puncture free.

Other packaging hints that will be helpful: place a double thickness of moisture-vapor proof paper between each steak or chop before you stack them in a single wrapping. This will enable the chops to separate easily for thawing. Put moisture-vapor proof paper between hamburger patties too. Stews and stocks freeze well in snap-lid plastic containers.

It is best not to fill the containers all the way to the top. Allow at least one-half inch of "headroom" for the liquid to expand. Label each package clearly. The label should indicate the kind and cut of meat, weight or number of servings, and the date on which it was frozen.

The dos and don'ts of freezer organization—To make your freezer work for you most efficiently, it is best to keep it well stocked and constantly draw on its supply. Do stack the same kinds of meat together. The dates on the packages will help you know how to rotate the meats in the freezer—using those first which have been frozen longest. Those frozen longest should be easiest to get to. Do freeze only the amount of a particular meat that you calculate can be consumed within the recommended storage period. Don't try to freeze too much at once. It will raise the interior temperature of your freezer. Do watch the newspapers for supermarket specials. Buy in quantity when prices are low and quality high. If freezer space is available and your budget will allow, take advantage of the savings and buy primal cuts, but be sure to read Chapter 19, "Buying Beef in Bulk—and the Chilling Truth About Freezer Plans," before you do. Finally, do keep a running account of what's inside your freezer. Like this:

KIND OF MEAT	AMOUNT	DATE FROZEN	REMOVED
Rib roast	1-six lbs.	8/14	√
	1-eight lbs.	10/2	
Pork chops	1 pkg. 4 chops	6/13	√
	1 pkg. 4 chops	7/16	
	1 pkg. 2 chops	8/14	
Hamburger	1 pkg. 4 patties	8/14	√
	1 pkg. 8 patties	10/2	
Leftover beef bone/scraps	1 pkg. (2 bones)	10/16	
Beef liver	1 container (4 servings)	9/16	
Lasagna casserole	8 servings	10/4	

Such an inventory will not only help you to plan your menus in advance—but also help you to come up with "instant" meals when unexpected guests arrive.

Some hazards of freezing—In most cases, if the rules are adhered to, freezing is a totally dependable method of storage. There are certain hazards, however, that one should be aware of:

- Fried meats will lose crispness in a freezer.
- Broiled meats may lose some of their original flavor.
- Bacon and seasoned sausages do not take well to freezing. The salt they contain speeds up the rancidity of their fat, and also stalls the freezing process.
- If meat is inadequately packaged, freezer burn—those innocuous spots of grayish white on the surface of the meat—is likely to occur.
- A loss of color, flavor, and nutritive value can also occur if the meat is inadequately packaged.

Cooking for (and from) the freezer—Frozen meats and meat dishes not only offer time and money-saving advantages, but they can go from the freezer to the oven, broiler, or range-top with ease. Cooking for the freezer can involve as much as an elaborate planned-in-advance company dinner, or as little as cooking in quantity. By doubling up on recipes, the freezer enables one to serve a dish today and reserve another for a later occasion. In any case, it is best to shorten the cooking time of a cooked meat dish you plan to freeze so that it will not be overcooked when reheated. Cooked meat dishes prepared for the freezer should be thoroughly cooled. If the dish prepared for freezing calls for white potatoes, add them while reheating the dish. The texture of potatoes tends to break down under freezing conditions.

Save leftovers in the freezer for a later date. Freeze bones and scraps of meat for stocks and soups. Store highly concentrated stocks in well-covered ice cube trays. Use individual cubes for flavoring or instant cups of bouillon.

Use freezer-to-oven cookware, if you have it, or aluminum foil pans for cooking meat dishes from the frozen state. Depending on the cooking method, hard frozen cuts of meat can usually be cooked as successfully—rare, medium, or well done—as corresponding cuts of fresh meat. The amount of cooked time required will vary, however, according to the cooking method employed.

Roasting frozen cuts requires at least one-third and up to half again the amount of time it takes to cook a corresponding fresh cut. The frozen roast can be placed in the oven preheated to the recommended temperature just as if it were fresh. After it has thawed in the oven and begun to cook, a meat thermometer inserted at the thickest part of the meat will help to determine the exact degree of doneness. Because it might break, do not attempt to insert the thermometer before the roast is thoroughly thawed. Roasting from the freezer should only be practiced when there is not enough time for you to thaw meat. As an overall practice, thawing prior to roasting is more satisfactory because there is a slight tendency for thawing ice crystals to steam the meat rather than just roast it.

Baking frozen fresh meat loaves will require about one-third more time than the amount recommended for unfrozen meat loaf. Frozen cooked meat casseroles should be baked until they are thawed and heated all the way through. Cover the top of casseroles loosely in foil to prevent them from burning.

The extra **broiling** time required for frozen steaks, chops, and hamburger patties will vary according to their thickness. Thin cuts require very little additional time, but the thicker cuts may need to be cooked twice as long. Thick frozen steaks should be placed on a rack at least four inches from the source of heat, or the broiling temperature should be reduced. This will permit the interior meat to reach the desired degree of doneness without burning the exterior surface of the meat. Test for doneness by inserting the tip of a knife in the center of the cut or close to the bone.

Frying, pan broiling, and sautéeing are methods of cooking that are much more successful with thawed meats because a certain amount of fat is required. As the ice crystals thaw, they release moisture. In combination with cooking fat, this moisture will cause an inordinate amount of splatter.

Moist cooking of frozen meats takes very little or no extra time than moist cooking fresh meats. Only the largest cuts will require slightly longer. If the meat is to be browned in fat prior to the moist-cooking process, however, it is best to thaw it first.

Thawing meat—Thawing can be accomplished either during the cooking, as previously described, in the refrigerator, or at room temperature. Depending upon the size, shape, and thickness of the cut, it will require from 5 to 8 hours per pound to thaw in the refrigerator, or approximately 2 hours per pound at room temperature. Meat can be thawed in its wrapping and used as soon as possible after it has thawed.

Here's a tip practiced by many restaurants to inhibit the loss of juices in thawing frozen meats at room temperature. Simply give all the outside surfaces of the meat a good coating of vegetable oil, and the juices will remain intact.

Although frozen meat will last for a few days in the refrigerator after it has thawed, it will lose some of its juices and its original color will darken. It is best to use thawed meat promptly.

MAXIMUM TIME RECOMMENDED
FOR STORAGE AND FREEZING

(To be effective a home freezer should have an interior temperature of no higher than ten degrees above zero. It is a good idea to keep a thermometer inside the freezer in order to be assured of this temperature.)

TYPE OF MEAT	NUMBER OF DAYS IN REFRIGERATOR	NUMBER OF MONTHS IN THE FREEZER
Beef		
steaks	3–4	6
roasts	4–5	8–12
cubed	3	6
chopped	3	3
corned	7	6
Pork		
chops	2–3	3
roasts	3	6
ground	1–2	3
cubed	1–2	3
spareribs	2–3	6
smoked	5–6	6
Ham		
whole, smoked	7	6
steaks	3–4	6
sliced, vacuum packed	7	*
canned	14–21 or more	*

* Not recommended for freezing.

TYPE OF MEAT	NUMBER OF DAYS IN REFRIGERATOR	NUMBER OF MONTHS IN THE FREEZER
Bacon		
sliced	7	*
slab	7–14	*
canned	14–21	*
Lamb		
roasts	3–4	6
chops	2–3	3–6
cubed	2	3
ground	1–2	3
Veal		
roasts	3–4	6
chops	2	3
scallops, cutlets	2	3
ground	1–2	3
cubed	1–2	3
Variety meat		
heart	2–3	3
liver	2	3
brains	2	3
kidney	2	3
tongue	2–4	3
sweetbreads	2	3
tripe	2–4	6
Sausages		
fresh	2–3	3
smoked	5–7	6
cooked	5–7	6
dry, semi-dry	10–14	6
Ready-to-eat luncheon meat		
sliced	2–3	*
unsliced	5–7	*
vacuum packed	7	*

* Not recommended for freezing.

TYPE OF MEAT	NUMBER OF DAYS IN REFRIGERATOR	NUMBER OF MONTHS IN THE FREEZER
Cooked meat dishes		
bones	7	3
leftovers	7	3
casserole	7	3
frozen dinners	–	3
stocks, gravies, sauces	7	6
soups, stews	7	6

◦╂◦╂◦╂┥

Chapter 22

The Gentle Art of Carving and Serving Meat

basic equipment... caring for it... advance preparations... carving techniques... serving tips

Great carving is great art. Good carving is a necessity to get the most tender and flavorful morsels of meat—and the least waste from any cut. Good carving also offers the bonus of good-looking leftovers or even extra meals; in other words greater consumer power. Carved and served with style, the most humble cuts can make for a memorable meal. In contrast, there is nothing that can spoil the style of a meal as much as a sloppy carver and unappetizing, hacked-up meat, piled unattractively on a platter. Yet anyone can master carving techniques. All you need is the proper equipment, practice, and a little patience.

The carving equipment you choose need not be fancy, but it must be strong, sharp, and efficient. When one is carving at the table, all eyes are on the perfection of the performance—and not on the staghorn-handled carving set. Better to invest in the quality of the blades, and spend time boning up on the proper techniques. Listed below are some basic pieces of carving equipment to set you on the way to receiving compliments on all the cuts of meat you serve.

THE SEVEN BASIC PIECES
OF CARVING EQUIPMENT

1. Nine-inch knife with curved blade, pointed tip
2. Eleven-inch knife with narrow, straight-edged blade
3. Two-tined fork with guard
4. A cloth napkin
5. A carving board equipped to catch juices
6. A knife steel
7. An emory stone

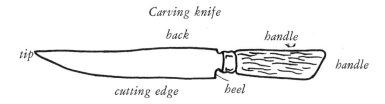

Carving knife

1. **The nine-inch knife** with a curved blade and pointed tip is the best general knife you can own for carving any cut of meat—particularly medium-sized roasts. The curved tip allows one to carve around a bone with ease. When selecting a knife of this type, insist on holding it before you buy it. The knife should have a certain heft to it, feel comfortable in your hand, and be well balanced and sharp.

2. **The eleven-inch knife** with a narrow, flexible, straight-edged blade and rounded tip is especially suitable for tackling whole hams, standing ribs, and other larger cuts. Its long, flexible blade cuts smoothly across the wide surfaces of meat, and facilitates the carving of thin, even, uniform slices.

3. **The two-tined** fork acts as a "steadying influence" for the carver, holding the cut firmly in place as he carves. The guard, at the heel of the blade, although not entirely necessary, prevents one's hand from slipping on the fork. The fork need not match the knife, although many carving knives and forks are sold as sets.

4. **The cloth napkin** (yes!) or a less pretty but more practical double thickness of paper toweling can give you the look of the professional when tackling cuts such as hams and legs of lamb. Many purists insist that the cut of meat should never be pierced with the

tines of a fork as it is being carved, their reason being that unnecessary juices will escape. For them—and maybe for you—the napkin is preferable. Although this method differs slightly from the one described in this chapter, it allows one to grasp the shank end of a cut firmly in one hand, while carving the meat with the other.

5. **The carving board** should be made of wood preferably. China is too slippery for successful carving. Silver or metal scratches. The board should have an indentation—the classic "well-and-tree" design is an example—for catching the juices of the meat.

6. **The knife steel** is one of the two pieces of equipment that's a must for the proper maintenance of your knives. The steel itself is generally about 8 inches long, either flat or round in shape with a handle, and works with magnetic action. The steel is sometimes sold as the third part of a three-piece carving set, but may be purchased separately also. The function of the steel is to maintain the sharpness of the knife blade, but it needs no special care itself. (One of the newer items on the market designed to replace the traditional steel is the ceramic honing stone. It works in the same way to maintain knife blades. Chances are you will hear more of it in the future.)

7. **The emory stone**, available in most hardware stores, is used to produce a sharp cutting edge on a knife blade. Many emory stones have a two-sided finish: a coarse side for sharpening the blade, and a fine side for smoothing it.

THREE OPTIONAL EXTRAS

Once you get the hang of carving you may want to expand your collection of equipment. Here are three of the accessories you can add:

The carver's helper is a fork-like tool with widely spaced tines designed for pinning down and steadying larger cuts of meat. The carver's helper can be used in conjunction with the long, slender, flexible-bladed knife.

The steak knife is a smaller version of the general 9-inch, curved-bladed carving knife. Its blade is generally about 7 inches long and it is often sold with an accompanying fork as a "steak set." This duo is ideal for carving poultry as well as steaks.

The electric knife is really not a knife at all, but a mini-power saw that may prove useful on certain occasions. It is not designed to replace the knife used under your own locomotion. It may come in handy when one is trying to work through bone or split a piece of poultry.

CARING FOR CARVING EQUIPMENT

Once you have made the investment in really good carving equipment, you can expect a lifetime of sharp service from it if you pamper it with proper cleaning, storage, and sharpening.

Steel-bladed carving knives and forks should not be soaked in hot, sudsy water. Not only can prolonged soaking loosen the bond between the handle and the blade, but it also tends to dull the luster of the shine. Proper cleaning only requires that you wipe each piece of equipment separately with a warm, moist cloth to remove the grease—then dry it.

Carving equipment can be stored in the box in which it was purchased if it has slots for holding the pieces firmly and protecting the cutting edge. Or you may want to invest in either of several kinds of knife racks—the sort that sit on a counter, hang on a wall, or fit snugly into a drawer. In any case, good carving equipment should not be allowed to rattle around in a drawer with other random pieces of kitchen equipment. Not only will this dull both the edge and the shine of the knife blades, but also it can be the potential cause of a nasty accident—cut fingers.

The wooden cutting board also has certain minimum requirements to keep it in good shape. The board should never be soaked in hot water, but merely wiped thoroughly with soap and water to remove the grease, rinsed, and thoroughly dried. If grease has built up on the surface of the board, the best way to remedy the situation is to cover its surface with a thin layer of salt to draw out the grease. Once a carving board starts to chip and splinter it has lost its usefulness. Get rid of it right away.

Aside from some of the newer "self-sharpening," serrated-edge carving knives, all knife blades tend to dull after prolonged use. That is why the knife steel and honing emory stone are basic pieces of equipment to the confirmed carver.

The knife steel should be used at the table prior to each carving assignment. "Steeling" the knife is a ceremony unto itself when done with confidence and flourish. You'll want to practice steeling privately before you try it for a crowd. Here's how:

How to steel a knife

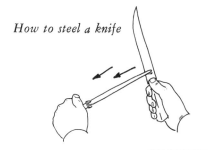

Step 1. Hold the handle of the steel in your left hand, the handle of the knife in your right. Place the heel of the knife at an angle at the tip of the steel, as illustrated.

Step 2. Holding the steel firmly, and maintaining the angle, draw the knife down to the base of the steel in one smooth stroke.

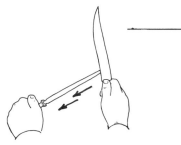

Step 3. For the other side of the knife, use the opposite side of the steel and repeat the procedure; maintaining the angle, draw the knife down to the base of the steel in one smooth stroke.

National Live Stock and Meat Board, Chicago

The steps described above help straighten and maintain both sides of a cutting edge of the knife blade. Six smart strokes on each side of the steel, or twelve strokes totally in quick succession, should be sufficient to maintain a sharp edge.

The emory stone need not be used as often as the steel. The stone serves to produce a sharp cutting edge, whereas the steel is used to maintain one. The emory is used before knives are brought to the table, in the privacy of one's own kitchen.

To use the stone, saturate it first with light oil—vegetable oil will do. (Some experts claim that soaking it in water works just as well. If you prefer oil, the stone will have to be thoroughly cleansed after each use, or stored by completely submerging it in oil. The use of water will eliminate that necessity.) The second step is to place the stone—rough side up—on a flat surface. Then proceed as illustrated in the following steps:

Honing a knife

Step 1. Hold the knife gently and place the blade at an angle to the stone, as illustrated. Work the knife gently across the entire surface of the stone from the tip to the heel—toward you.

Step 2. Turn the knife over and reverse the procedure, working away from you.

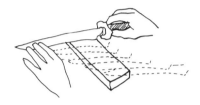

About four strokes across the stone on each side of the knife blade on both the rough and the fine sides of the stone's surface should be sufficient to produce a sharp edge. Test for sharpness by pressing a finger gently on the cutting edge of the knife.

One warning: if you care for your knives and want the maximum use from them, *never* attempt to sharpen their blades on an electric knife sharpener. The electric sharpener will actually tear the steel and drastically shorten the life span of your knife blade.

ADVANCE PREPARATIONS

Elbow room is essential to carving, so make sure you have plenty of it. Dispose of stemware, candlesticks, vases or other non-essentials which might hinder the carver's progress, and allow enough space to accommodate the platter of meat to be carved, the serving platter or individual plates, and the carver's utensils. If the dining area is small, the actual carving might best be accomplished at a sideboard or table.

In advance of carving, allow sufficient time for the cut of meat to "set" after it has been removed from the oven. Perhaps the best way to accomplish this and still keep the meat warm is to simply turn off the oven and let it sit there. If you must plan other items on your menu around this resting period and need the oven space, an alternative method for retaining the heat is to construct a tent of aluminum foil large enough to completely cover the cut.

According to the size of the cut and the degree of doneness, you should allow the meat to wait in a warm place from 15 to 30 minutes before carving. This allows the juices to retreat and the meat to become firm enough to make the task of carving somewhat simpler. A rare, large cut of meat requires up to 30 minutes to set, while well-done meat requires only half that. The exception is steaks and small individual cuts of meat which are always served immediately after cooking. Remove any excess strings or skewers in the kitchen except those on a rolled or stuffed roast which are removed as the carver proceeds.

Before embarking, also make certain that the cut of meat is pointed in the proper direction from which to carve it. Over garnishing on the carving board is a definite no-no. It is perfectly permissible for the carver to deftly dispose of a flamboyant spray of parsley or watercress by removing it to the serving platter with the tip of his knife and the fork. Vegetables, likewise, almost never share the same platter with meat to be carved—regardless of the pretty pictures one is apt to see in advertisements.

The advance preparations are now complete. You are now ready to stand up (or sit down—whichever is most comfortable to you), sink your fork in the proper position, raise the knife, and begin . . .

A GUIDE TO CARVING TECHNIQUES

The cardinal rule of carving is: cut across the grain—not parallel to the fibers of the meat. "Grain" is the fibrous texture of the meat, clearly visible in a cut such as beef brisket. Cutting in the same direction as the grain can make even the best meat stringy. Whereas cutting across the grain helps to ensure tenderness, as does thin slices. There may be some rare instances when it is more practical to violate the rule, but on the whole it is best to stick with it.

For the most uniform and attractive slices, get a good firm grip on the knife and keep its angle constant. Use long, sweeping strokes rather than choppy, sawing motions. Keep the meat steady by holding the fork behind the knife and puncture the meat with the fork as little as possible to prevent the juices from escaping. Serve the slices with both knife and fork.

The bone structure of a cut will govern the actual pattern of procedure. That's why a little knowledge of animal anatomy certainly can't hurt your carving. Boneless cuts are, of course, the easiest to carve

—bone-in shoulder among the most difficult. Similar methods of carving can be applied, however, to similar cuts such as a leg of lamb and a whole ham. The methods illustrated below are carving guidelines for the major cuts of meat. They begin with the simplest and graduate to the more complex. All of these directions are written and illustrated for the right-handed carver—so if you're a lefty you will have to make the proper adjustments.

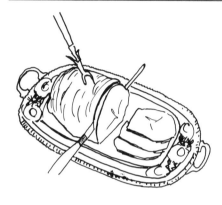

Rolled roasts (pork, lamb, veal, and beef) can be devised from any section of meat. They are generally from the shoulder, leg, round, or chuck. The general method of carving requires that one slice down vertically from the top of the roast to the cutting surface. Remove the strings only as you come to them to prevent the roast from falling apart.

Rib-eye roast (beef) is carved almost identically to any rolled roast except that the knife is held at a slight angle (as illustrated) to help keep the juices in the meat.

Stuffed roasts (pork, lamb, beef, veal)—Carving a stuffed boneless roast shoulder or breast couldn't be simpler. The slices are simply vertical from the top of the roast to the cutting board. Two hints that may be helpful: do not remove any strings until you reach them, and serve the slices with the help of a spatula to keep the stuffing intact.

National Live Stock and Meat Board, Chicago

Brisket and corned beef at its tenderest best must be ultra-thinly sliced. Before proceeding to carve, remove any excess fat. Then, starting at the round end, cut slanting diagonal slices from three planes in rotation. Alternate the slices from plane to plane in order to keep their size constant as you proceed to carve down the brisket.

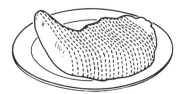

Tongue is another fibrous cut that should be thinly sliced for maximum tenderness. Place the tongue on its back with the tip to the left and slice downward and at a slight angle as illustrated.

Crown roasts (pork or lamb) devised of rib chops of lamb or pork may look complicated to carve, but they are deceptively easy. You simply divide the spectacular-looking crown into separate chops by cutting downward vertically between each of the ribs. Remove each chop as it is carved and allow two chops per serving.

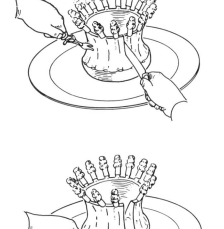

National Live Stock and Meat Board, Chicago

Loin of pork—Carving a loin of pork is a simple, two-step procedure if the rib bones are sawed through to separate them from the backbone. This is the way a pork loin is usually prepared before it is wrapped at the supermarket. Unless this is accomplished, however, the task of carving is difficult at best, and at worst practically impossible. Before you buy it, make certain the ribs have been sawed.

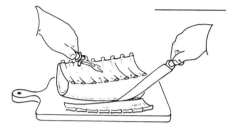

Step 1. Remove the backbone, as illustrated. Insert the tip of your knife between the backbone and the meat and run it around and over the backbone, detaching it as cleanly as possible.

Step 2. Place the meat on its broad end, with the tips of the rib bones facing you. Slice downward vertically between each rib, allowing one chop per serving. For thinner slices and servings half that big, cut in as close to the bone as possible and along its side, alternating boneless slices with those that contain a bone.

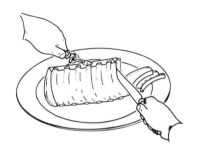

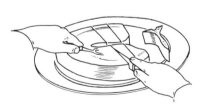

Porterhouse steak—Carving a Porterhouse steak—as well as sirloin or T-bone—is simply a matter of separating the meat from the bone by running the tip of a sharp knife around and as close to the bone as possible. The bone is then removed (as illustrated). To further dissect the steak, hold your knife at an angle, and beginning at the wide end,

National Live Stock and Meat Board, Chicago

cut across it in diagonal one-inch slices. The slant of the slice helps keep the juices in the steak. The tip—or tail—of the steak is more fibrous and not as tender as the section nearer the bone, so cut it across as illustrated to ensure servings with shorter fibers. As you serve, make certain that every plate gets a section of meat from the area near the bone, as well as one from the less tender regions.

Leg of lamb is one of the more complicated cuts to carve and may require some practice on your part before performing in the presence of company.

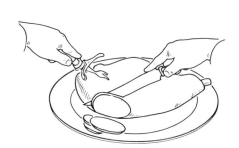

Step 1. Remove a few lengthwise slices from the underside of the roast in order to form a broad base on which to rest it. Then turn the lamb on the base. If it is a left leg (like the one in the illustration), the protruding leg bone will be to your right. If it is a right leg, the bone will be to your left. This is an opportunity for you to steady the roast by grasping the shank end in your hand with the use of the napkin if you wish.

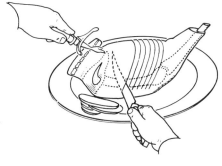

Step 2. Assuming the leg is a left one, and the bone is to your right, start at a point nearer the protruding shank bone and carve downward to the bone and toward you. (For a right leg slice downward and away from you.) Make thin parallel slices.

National Live Stock and Meat Board, Chicago

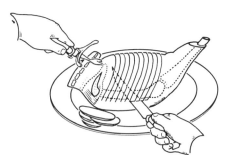

Step 3. After you have completed slicing the most choice center section, remove the slices of lamb from the bone by running your knife underneath them. Serve these slices first, but if someone wants seconds, the easiest way to carve the meat remaining on the bone is to remove it from the bone entirely.

Ham—A whole baked ham—either fresh or smoked—is carved by a method similar to that for leg of lamb.

Step 1. The shank end is always placed to the right of the carver. The lengthwise initial slices are then removed from the thin side of the ham in order to form a base on which it can sit. If the ham is a right leg of pork, the lengthwise slices are removed from the side farthest away. (An anatomical tip: differentiate between a right and a left ham by the bony kneecap. A right leg contains the kneecap on the side nearest the carver. A left leg contains the kneecap on the side farthest away.)

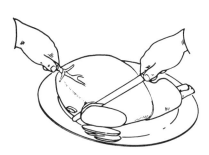

Step 2. Carve a parallel series of slices perpendicular to the length of the leg bone.

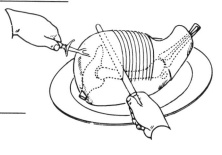

National Live Stock and Meat Board, Chicago

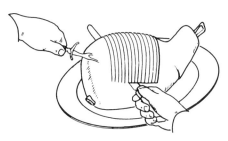

Step 3. All the slices can then be removed at one time— neatly and efficiently. Just run the knife underneath them along the bone. If more slices are desired, or a second round of helpings, re- move additional lengthwise slices from the base.

(Don't forget to save the scraps and bone for the basis of another meal. See Chapter 8, "Them Bones and the Gold Mine in Your Garbage Pail: Stock.")

Ham (shank half)—Carving the shank half of ham is a fairly simple affair because the only interior bone is clearly visible on the cut side.

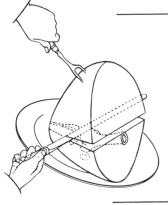

Step 1. Detach the thick upper portion of the ham by run- ning the knife along the top of the bone, as illustrated.

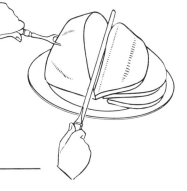

Step 2. Slice it, freshly-cut side down, at the side of your cutting surface.

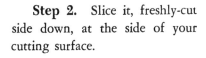

National Live Stock and Meat Board, Chicago

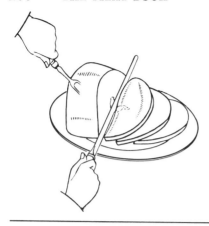

Step 3. The lower portion of the shank may then be sliced by removing the bone entirely with the tip of the knife. The now boneless piece is then inverted, bottomside up, and sliced in the same method employed when slicing the upper portion.

Chuck blade pot roast—An unboned chuck blade pot roast has natural dividing lines of bone and connective tissue that are easily visible. It is these patterns that your knife should follow when carving the blade roast in order to extract the largest solid sections of meat.

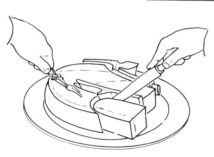

Chuck blade pot roast

Step 1. Remove all the bones with your knife.

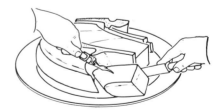

Step 2. Following the illustration, fork each section of meat perpendicular to the cutting surface.

National Live Stock and Meat Board, Chicago

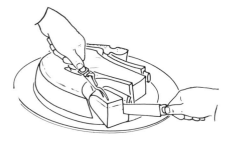

Step 3. Slice downward against the grain of the meat into sections one-quarter to one-third inch thick. This cut may also be purchased boned, rolled, and tied for pot roasting—which would certainly make a carver's life easier.

Beef rib roast—A standing rib roast is not as intimidating to carve as it looks. The roast is placed face down with the rib bones to the carver's left, the meat side to the right.

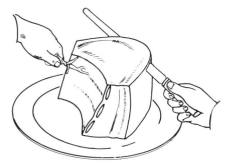

Step 1. Stick the fork between the ribs to hold the cut firmly, and slice straight across from right to left—horizontally.

Beef rib roast

Step 2. Free each slice from the bone as it is carved by running the tip of the knife along the edge of the rib bones.

National Live Stock and Meat Board, Chicago

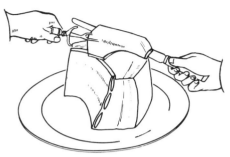

Step 3. Lift the slice to the serving platter by sliding the knife under it and pinning it down with the fork. One slice is an ample portion. Continue slicing horizontally until you have as many slices as you need. Slices may be ultra-thin or up to one-half inch thick, depending upon your preference.

Shoulder of pork (or smoked picnic ham) is complicated to carve because of its interior bone structure and will require considerable patience and practice to master. If the effort seems more than you can master, it is suggested that you purchase this cut de-boned. If, however, you want to try your hand at it, here's how:

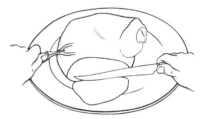

Step 1. Remove a few lengthwise slices, as illustrated, from the bottom—or smaller side —of the roast to form a steady base on which it will rest.

Step 2. Turn the roast over on its base. Find the point on top of the shoulder where the elbow ends, and slice down just in front of it. Then, turn the knife blade to the left, and slice along the length of the arm bone to its end, freeing the entire upper section of the meat from the bone.

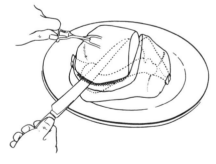

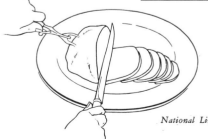

Step 3. Slice this boneless section of meat separately on the side into a series of perpendicular, equally thin slices.

National Live Stock and Meat Board, Chicago

Step 4. The remaining meat is then removed with the tip of the knife from both sides of the arm bone. These two rather randomly shaped boneless pieces of shoulder can be sliced thinly—but the slices will not be as uniform in size nor as attractive as the original slices.

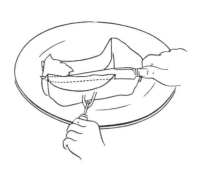

Butt half of ham has an interior bone structure that is quite complex and difficult for the inexperienced carver. It is a far greater challenge to carve the butt end of a ham than it is the shank half. To dissect the butt proceed as follows:

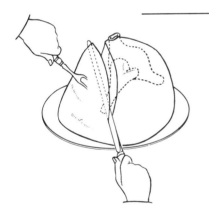

Step 1. Put the butt on the cutting surface, cut side down. Remove a large boneless piece of meat from the side of the ham by slicing down from the top, along the interior bone, all the way to the cutting board.

Step 2. Place the boneless piece of butt on its freshly cut side and carve it into thin, cross-grain slices.

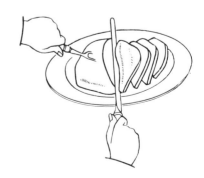

National Live Stock and Meat Board, Chicago

Step 3. The meat remaining on the other side of the butt can be carved like this: steady the roast with the fork and, using the knife, slice horizontally inward to the bone from right to left. Detach each slice from the bone with the sharp tip of the knife. Remove each slice as you carve it.

National Live Stock and Meat Board, Chicago

SERVING TIPS

As you carve and serve, there are a few simple and sensible tactics you'll want to observe. It is hardly necessary to mention that neatness does count. But it is important, for instance, to prevent the juices of the meat from escaping as much as it is possible. In order to do so, try to keep the number of times you puncture the cut of meat with the fork to a minimum. Carve all the meat from a single side of a roast before embarking on another, and carve only the amount you will need to serve everyone at the table. Meat dries out more quickly after it has been cut, so if there is enough meat left for second helpings, carve them later as they are needed.

Have warmed plates at hand to receive carved meat. Cool plates cause the juices to rapidly congeal, and make meat chillingly unappetizing.

If the beef or lamb you are carving is cooked medium or rare, reserve the first outer slices for people who prefer their meat on the well-done side. Most cuts will have an area containing the superior meat. Make certain to distribute the superior slices equally, allowing a slice from the choice area per plate, as well as one from the lesser regions.

Finally, if you are serving each plate as you carve, try not to be overly generous with the first few and then have to skimp on the last. Try to serve equal portions as much as you can, so that no one—especially the carver—gets left out. Everything else will be forgiven.

Appendices

1

AN ILLUSTRATED GLOSSARY
OF MEAT COOKERY TERMS

Au jus   A French term that literally translates as "in the juice." A meat dish served with its natural juices or gravy.

Bake   To cook by dry heat in an uncovered pan in the oven. When this cooking method is applied to meat, it is usually called roasting.

Barbecue   To roast meat slowly on a rack or revolving spit over or under the source of cooking heat. Charcoal, wood, oil, or gas can be used as fuel. During the process, the meat is often basted with a highly seasoned sauce.

Bard   To add extra fat (usually pork fat or bacon) to the exterior surface of lean cuts of meat. Barding gives meat extra juiciness and flavor, and prevents it from drying out during the cooking process.

Baste   To apply liquid with a spoon or bulb baster during the cooking process. Basting adds flavor to meat and prevents it from drying out. Lean cuts particularly benefit from the additional moisture provided by the fatty pan drippings or other basting liquid.

Batter   A liquid mixture usually composed of milk or water, flour, and eggs into which raw meats are dipped to help them form a crisp outer crust as they cook.

Blanch   To render meats such as tongue, ham, bacon, and salt pork of their dominant flavors. Meats to be blanched are placed in a kettle of cold water, brought to a boil, simmered according to recipe, then plunged directly into cold water to stop the cooking immediately and help to firm the meat.

Boil   To cook in bubbling liquid at 212° F. *Note:* Meat should be simmered, not boiled.

Bouillon   A strong meat stock that has been strained, clarified and degreased.

Braise
American Meat Institute

Braise   A method of meat cookery by which the meat is usually browned first in a small amount of fat, and then slowly cooked in a covered pot to which a small amount of liquid has been added. Meats may be braised over direct heat on top of the stove, or in the oven.

Bread   To coat meat in cracker, cereal, or bread crumbs prior to cooking. The meat may have first been dipped into a batter of lightly beaten eggs and liquid.

Brine   A salt and water solution used for preserving meats.

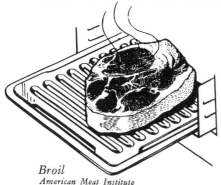

Broil
American Meat Institute

Broil   To expose meat on a broiler rack to a direct heat cooking method utilizing a temperature of 550° F. or slightly higher. It is usually advisable to place the broiler rack three to four inches from the source of heat.

Brown   To produce a dark coloration on the exterior surface of meat by searing it in fat before proceeding to cook. Meats browned slowly will shrink less than those that are quickly browned.

Casserole   A heatproof dish in which food is cooked, or the food served from such a dish.

Carve   To slice meat with a knife prior to serving it. See Chapter 22, "The Gentle Art of Carving and Serving Meat."

Consommé   A clear broth made by simmering meat and bones in a large quantity of water with vegetables, herbs, and seasonings. The vegetables most commonly included are carrots, celery, and onions. After cooking, the consommé is strained, degreased, clarified, and the seasoning is corrected.

Cracklings Crispy pieces of rendered pork fat.

Cube Small, uniform-sized pieces of meat, or to cut in small, uniform pieces.

Deep fry To cook meat rapidly in a large quantity of fat. The fat used must be very hot and deep enough to completely cover the meat to be fried. This method is very tricky and takes practice to perfect.

Deglaze To remove the extra fat from the pan drippings and meat sediments in which a cut was cooked, then making a pan gravy by adding stock or wine.

Dredge To coat meat completely in flour, cornmeal, breadcrumbs, or other fine-textured substance prior to cooking.

Drippings The fats rendered by cooking meat. Recommended for making tasty meat gravies. Pork and bacon drippings can be stored and used to flavor other foods.

En brochette A French term describing food cooked and served on skewers.

Fillet A boneless cut of meat, or to remove the bone from meat.

Fondue Raw cubes of meat cooked in hot oil and dipped into a selection of seasoned sauces.

Freeze To store meats at temperatures of less than 32° F. In order to be effective, the home freezer should maintain a temperature of no higher than ten degrees above zero. See Chapter 21, "Storing and Freezing Meats."

French fry To fry in deep fat.

Fry To cook meat over direct heat in a pan or skillet with or without the addition of extra fats. It is possible to pan broil foods in the new non-stick cookware.

Garnish A decorative or flavor-enhancing edible. Parsley and watercress are common decorative garnishes for meat.

Glaze An edible outer coating, sometimes sweet, or a meat stock reduced to a thin paste which is used to enhance other recipes.

Grease To coat a pan with butter or other shortening. Also, fats rendered by meats during the cooking process.

Griddle broil To cook meat uncovered on a griddle, scraping off the fat as it accumulates.

Grind   To change the textural composition of meat by reducing it to tiny pieces.

Kabobs   Cubes of meat threaded on a skewer and sometimes alternated with vegetables. Kabobs may be marinated, then grilled or broiled.

Lard   The rendered fat of hogs. Also the process of inserting pieces of pork fat into meat to keep it moist as it cooks.

Lardoon   The name given to the pieces of pork fat inserted during the larding process.

Marbling   The delicate interior webbing of fat within lean meat. Marbling is particularly evident in Prime and Choice beef.

Marinade   An aromatic, acidic liquid which works with enzymatic action to tenderize lean cuts of raw meat. Some marinades are used in the finishing sauce.

Mince   Tiny pieces of meat. Also to cut meat with a knife into fine, tiny pieces.

Pan broil   To cook meat in an uncovered skillet. Fat may be removed from the pan as it accumulates.

Pan broil
American Meat Institute

Pan fry   To cook meat with a small amount of fat in a skillet.

Parboil   A preliminary cooking method by which meats are simmered for a short time to expedite the final cooking method or to firm their texture.

Paté   A paste or spread made of one or a combination of meats (liver, pork, etc.). Sometimes served baked in a pastry crust.

Poach   See "Simmer."

Pot roast   A larger cut of meat that is usually browned first, then braised in a covered pot with a little liquid.

Precook   A preliminary method. The term is usually applied to meats rather than the term "parboil."

Ragout   A highly seasoned meat stew, with or without vegetables.

Rare   Underdone or, when applied to meat, deep pink.

Reduce   To concentrate the flavor of a meat liquid by boiling it rapidly to evaporate the water.

Roast
American Meat Institute

Roast   A chunky piece of meat, usually cooked by dry heat (see "Bake"). Also to bake by dry heat, usually in an oven.

Rotisserie   A device for cooking large, chunky cuts such as a leg of lamb. Meat cooked in this fashion is threaded on a skewer and rotated under or over the source of heat.

Sauté   To cook meat rapidly in an open frying pan. Thinly cut or diced meats are sometimes cooked by this method by agitating them rapidly in hot oil and/or butter. The sediments that accumulate in the pan are usually deglazed with stock or wine to make a sauce.

Scallop   An ultra-thin, flat cut of meat.

Score   To make shallow, even marks in a cut of meat, usually to prevent it from curling as it cooks.

Sear   To brown meat and seal it in its juices by a short, intense application of heat.

Simmer   To cook meat slowly in slightly stirring liquid at a temperature of approximately 185° F. This slow-cooking technique prevents shrinkage.

Skewer   A long, slender metal needle on which meat is threaded to hold it intact as it is cooked.

Spit   A revolving bar which is inserted through meat which is to be cooked over an open fire.

Steam   To apply moist heat to meat.

Stew
American Meat Institute

Stew   To simmer cubed meat in liquid to cover, with or without browning it first. Also, the dish made by long, slow moist cooking, usually made with vegetables and seasonings.

Stock   A rich, nutritious broth that serves as the basis for soups, gravies, and sauces. Made by simmering meat scraps, bones, vegetables, and trimmings in water over a prolonged period of time.

Suet   A thick lobe of beef fat from the upper part of the flank. Sometimes ground or diced for flavoring purposes.

Tartare   Double-ground, extra-lean beef usually eaten raw.

Tenderize   To render meat tender by breaking down the fibers during cooking, marination, or pounding.

❖❖❖❖❖

2

THE BASIC PRINCIPLES
OF ROASTING AND BROILING

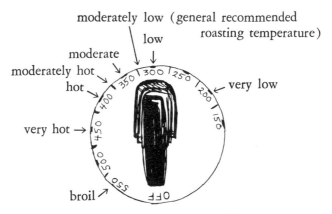

moderately low (general recommended
roasting temperature)
low
moderate
moderately hot
hot
very low
very hot →
broil
off

A Standard Dial for Regulating Oven Temperature

BASIC ROASTING PRINCIPLES

Equipment essentials:

A shallow, lidless pan large enough to accommodate the roast comfortably.

A rack to fit in the pan and prevent the meat from literally "stewing in its own juices." (Some loin and rib cuts form a natural rack which will hold the meat above its drippings.)

A meat thermometer to take the guesswork out of roasting and determine the exact degree of doneness. Because the weight, size, shape, and thickness of roasting cuts is extremely variable, roasting can be a very tricky cooking method. Charts which offer recommended cooking times according to weight are not totally reliable. A good meat thermometer is a very sound investment if you want the roast you serve cooked to perfection.

Method #1: Non-searing, recommended for less shrinkage of the meat.

Steps:

A.   Remove the meat to be roasted from the refrigerator at least one hour before you plan to pop it in the oven. Large cuts should be taken out even earlier as they will require more time to reach room temperature.

B.   Preheat oven to recommended temperature—usually 325° F.—although recipes sometimes differ.

C.   Season the roast if desired, following your favorite recipe. Lean cuts such as sirloin tip will ultimately be juicier if a little extra oil is rubbed on the outside surface.

D.   Place meat on rack in pan, fat side up. Insert meat thermometer in the thickest part of the roast without the tip of the thermometer touching fat or bone. Put the roast in the preheated oven.

E.   Do not cover, add liquid, or baste. Roast until the thermometer registers five or ten degrees below the desired degree of doneness. Remove the roast from the oven.

F.   Allow the roast to "set" for 15 or 20 minutes before carving. It will continue to cook slightly after it is removed from the oven, which accounts for why it was removed slightly "underdone." During the setting period the juices will retreat into the roast and it will become easier to carve.

Method #2: Searing—recommended for richer, browner exterior and drippings of beef and lamb roasts only.

Steps:

A.   Same as Step A, Method #1.

B.   Preheat the oven to 450° F.

C.   Same as Step C, Method #1.

D.   Place meat on rack in oven, fat side up. Insert meat thermometer in the thickest part of the roast without the tip of the thermometer touching fat or bone. Put in a 450° F. oven for 20 to 25 minutes. Do not cover, add liquid, or baste.

E.   Reduce oven temperature to 300° F. Allow oven temperature to cool slightly by opening door of the oven for a few minutes.

F.   Close the door to the oven. Roast until the meat thermometer registers five or ten degrees below the desired degree of doneness. Remove the roast from the oven.

G.   Same as step F, Method #1.

ROASTING TIMETABLE:
ESTIMATED TIME PER POUND FOR
DESIRED INTERNAL TEMPERATURE
(Constant oven temperature: 325° F.)

	DESIRED THERMOMETER READING:	APPROXIMATE MINUTES PER POUND:*
Beef		
Rare	125	13
Medium-rare	140	15
Medium	160	18
Well-done	170	20 to 25
Pork**		
Fresh	170	35
Smoked	170	35
Ham		
Cook before eating	160	30
Fully cooked	–	15
Lamb		
Pink	135	12 to 15
Well-done	165	30 to 35
Veal**	170	35

* Add 5 to 10 minutes per pound for boneless, rolled roasts.
** Pork and veal are always cooked well done.

SOME SUGGESTED CUTS
FOR ROASTING

Beef

rib roast   eye round
top sirloin butt   ground beef loaf
top round   beef tenderloin
top sirloin   strip loin roast
chuck shoulder

Ham (Smoked Pork)

ham (cook before eating)   smoked pork shoulder
ham (fully cooked)   smoked tenderloin
smoked pork loin

Pork (Fresh)

center cut loin   shoulder
center cut rib   butt
full loin   spareribs
fresh ham (leg)

Lamb

whole leg   shoulder
shank half of leg   rack of lamb
sirloin half of leg   loin of lamb

Veal

leg   rolled shoulder
rib (rack)   loin

BASIC BROILING PRINCIPLES

Equipment essentials:

A broiler pan with rack, preferably the kind that fits in an adjustable broiler so that one can vary the distance of the meat from the source of heat. Meats a quarter-inch thick should be about 2 inches from the direct heat; meat 1 inch thick should be 3 inches from heat, 1½ inch thick meat 4 inches; 2 inch thick meat 5 inches. Since the actual broiler heat can vary, these recommended distances are approximate.

Spatula or tongs for turning meat.

Method:

A. Select tender, high-quality, relatively thick cuts of meat for broiling. Pan broil thinner cuts. Steaks, chops, and patties should be at least a quarter of an inch thick. Ham slices at least a half inch thick.

B. To hasten cooking time, take the meat to be broiled from the refrigerator an hour before cooking. Preheat the broiler to broiling temperature. This is usually the hottest temperature on the oven dial. Slash the fat of the steak or chops in several places to prevent it from curling as it cooks.

C. Place the meat on the rack in the broiler at the appropriate distance from the source of heat according to the thickness of the cut.

D. Broil the meat until brown on top (ham slices need only be lightly browned). Turn the meat without pricking the exterior, using tongs or a spatula. Brown the opposite side of the meat. To test for doneness: insert the tip of a small knife close to the bone or the center of the cut. You can determine if it is cooked to your taste by the interior color.

E. Remove the meat from the broiler. Season if desired. Serve immediately on a heated platter.

BROILING TIMETABLE

	APPROXIMATE MINUTES PER SIDE*		
	RARE	MEDIUM	WELL DONE
Beef			
Steaks 1″ thick	5	7	10
1½″ thick	9	11	13
2″ thick	15	18	21
Patties ¾″ thick		5	
Filet mignon 1″ thick	3	4	5
1½″ thick	4	5	6
2″ thick	5	7	9

*Cooking time is gauged for meats at room temperature. Allow more time for chilled meats fresh from the refrigerator.

Lamb

Chops and steaks 1″ thick	4	7	8
1½″ thick	7	9	11
2″ thick	11	13	18

Pork

Smoked ham sliced

(fully cooked) ½″ thick	–	–	5
1″ thick	–	–	8
bacon	Until crisp according to personal preference		

SOME SUGGESTED CUTS
FOR BROILING

Beef

Porterhouse steak   strip loin
rib steak   Delmonico
sirloin (hip) steak   rib eye
top round steak   top butt sirloin
chuck steak   chuck tender
filet mignon   top round
ground beef patties   kabobs

Pork*

(smoked)   bacon
ham slice

Lamb

loin chops   sirloin chops
rib chops   blade chops
lamb patties   arm chops
kabobs   lamb steaks

Veal*

*Pork and veal are generally not recommended for broiling. The cuts adapt better to braising or sautéeing.

3

BEEF, LAMB, PORK, HAM AND VEAL
CUTS EXPLAINED—AN ILLUSTRATED GLOSSARY
OF THE RETAIL CUTS

Names vary from one geographical area to another and are constantly changing and being newly created. Some states require that the primal cut also appear on the label if one of the more fanciful names is used for a particular cut.

NAME   DEFINITION

BEEF

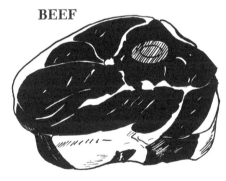

arm roast   Roast from the chuck section, sold either boneless or bone in. Roast or moist cook.

arm steak   A steak from the chuck with or without a bone. Pan fry if thin. Moist cook if thick.

barbecue   A boneless or semi-boneless chuck steak. Pan broil or barbecue.

bastard steak   The spare ends of almost any cut of steak. Pan broil or fry.

beauty   First-cut rib steaks with the bone removed. Broil.

beef filet   A tender, boneless steak cut from the loin. Also called a filet mignon. Broil or pan broil.

beef kabob   Small, lean boneless cubes of beef. Can be cut from round, chuck, or loin sections. Broil or braise.

beef rib   Any part of the rib primal cut with the bones intact. Roast or broil.

beef stroganoff   Boneless, ultra-thin slices of sirloin, served in a sour cream sauce.

beef sukiyaki   Paper-thin slices of sirloin served sautéed Oriental style.

bell of knuckle   A roast from the dry, lean side of the round. Moist cook.

blade cut pot roast   A flat chuck roast that must be moist cooked.

blade steak   A less tender chuck steak. Pan fry if thin. Moist cook if thick.

boiling meat   A fatty cut from the plate. Moist cook.

bone-in brisket   A fatty cut from the breast of beef. Requires long, slow moist cooking.

boneless club   An individual steak from the top of the short loin. Broil or pan broil.

boneless cross-cut shank   A sinewy cut from the foreshank suitable for stew or stock.

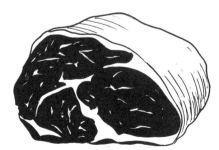

boneless rib   Also known as Delmonico, a no-waste rib roast or steak. Broil the steak. Roast the roast.

boneless riblifter   A tender, end-cut rib steak. Broil or pan broil.

boneless rump   A lean roast from the hip end of the round. Pot roast or moist cook.

boneless shoulder   A lean, meaty chuck steak or roast. Best braised or moist cooked.

boneless sirloin   Any steak cut from the sirloin without the bone. Broil.

bolar   A very lean chuck roast. Moist cook or roast with added fat.

Boston cut   A blocky, bone-in chuck roast. Moist cook.

bottom butt   A cut of sirloin steak nearest the round. Pan fry.

bottom round   A rather dry roast or steak from the round. Pot roast or braise.

bottom sirloin   End roast from the sirloin nearest the round. Roast.

bottom sirloin butt   A roast from the end of the sirloin nearest the round. Roast.

braciola   Very lean, Italian-style rolled round roast. Braise or moist cook.

bread and butter   Best of all the center-cut chuck steaks. Pan fry.

breakfast   Thin, lean chuck steaks. Pan fry.

brisket   A boneless roast cut from the breast of beef. Pot roast.

bucket   Oval-shaped steak from the side of round. Broil.

butter   Small, grainy chuck steak. Fry or pan broil.

California   Semi-boneless, center-cut chuck roast. Pot roast.

capital cut   A bone-in rib steak. Broil or pan broil.

catfish tender   A front-cut chuck roast. Moist cook.

center cut   A tender, very lean round steak. Pan fry or braise.

center-cut club   A well-marbled, tender rib steak. Broil.

chateaubriand   A superb roast from the beef tenderloin. Also the steak known as filet mignon. Roast or broil.

chicken   A small chuck steak with a streak of gristle in its center. Pan fry.

chicken fry   A steak cut from the top of the round. Pan broil or fry.

chuck fillet   Roast cut from the center of the chuck. Roast or moist cook.

chuck tender A lean chuck steak with a streak of gristle or cartilage in the center or a chuck roast with the cartilage removed. Roast with added fat or pot roast. Pan broil the steak.

chuck wagon Boneless, round-shaped chuck steak. Pan fry.

clod A lean, boneless cut of chuck—roast or steak. Pot roast or pan fry.

club A steak from the rib end of short loin. Broil.

club rib A semi-boned rib steak or bone-in rib roast. Broil the steak. Roast the roast.

corned beef Cured brisket from the breast of beef. Moist cook.

crescent A lean, crescent-shaped round roast or steak. Roast the roast. Pan fry the steak.

cross cut shank A bone-in or boneless cut of stew meat from the shank.

cross rib A bone-in chuck roast. Pot roast or roast.

cube Small, individual-portion steaks of chuck round or flank that have been machine processed to break down the elastic connective tissue they contain. Pan broil.

culotte A loin end (sirloin) steak. Broil or pan fry.

Dallas cut A chuck roast with fat and bone removed. Pot roast.

Delmonico A filet of the rib of beef, both a steak and a roast. Also a steak from the rib end of the loin. Broil or roast.

Denver pot roast Well-trimmed heel of the round. Moist cook.

diamond Very lean roast from the side of the round. Roast.

double-bone sirloin A wasty, first-cut or pinbone sirloin steak. Broil.

easy-to-carve rib A semi-boned rib roast. Roast.

English cut A roast from the chuck with the rib bones intact. Moist cook.

eye A boneless, elongated roast from the round. Roast.

face A steak from the side of the round. Pan broil.

face round A roast from the side of the round. Roast.

filet Whole tenderloin roast. Roast.

filet mignon Tenderloin steak. Broil.

first cut The best cut from either the sirloin or short loin. Broil.

flank steak filet A London broil trimmed of waste. Braise.

flanken beef Short ribs for moist cooking.

flat A square, flat, boneless chuck roast. Pot roast.

flatiron A boneless chuck roast that is divided in the middle. Roast or broil.

Frenched rib Rib roast with the fat of the rib bone removed. Roast.

full cut round Very lean steaks of top round. Pan broil.

gooseneck The whole bottom round trimmed for roasting. Also a steak from the bottom round. Broil or pan broil.

ground beef Any section of the beef that has been ground. Broil or pan broil.

half cut A top round steak that has been cut into two smaller halves. Broil.

hanging tenderloin A tough steak from the loin and kidney section. Braise.

hip The pin bone sirloin steak cut from the hip section. Broil.

hip bone Steak cut from the center of the loin sirloin. Broil.

his and hers Any large steak suitable for two people. Broil.

horseshoe Heel end of the round. Pot roast or moist cook.

hotel A bone-in shell steak or boneless strip steak from the top of the loin. Broil.

inside chuck roll Boneless, rolled center-cut chuck roast. Pot roast.

inside round Top of the round roast cut into either a steak or roast. Pan broil the steak, roast the roast.

inside top round The same cut as the inside round—the best part of the round for steaks and roasts. Pan broil or roast.

Jew daube   A chuck roast. Moist cook or pot roast.

Jewish filet   A boneless, center-cut chuck roast. Pot roast or roast with added fat.

Jewish tender   Soft part of the chuck suitable for both steaks and roasts. Pan broil or pot roast.

jiffy   Thinly cut or cubed flank steak. Pan fry.

Kansas City   Well-trimmed Porterhouse or T-bone steak. Broil.

Kansas City broil   A bone-in or boneless, center-cut chuck steak. Pan broil.

king   A steak from the large end of the short loin (Porterhouse). Broil.

knuckle   The lean side of round for roasts and steaks. Pan broil the steaks. Roast the roasts.

large T-bone   Steaks from the short loin. Same as the Porterhouse with a smaller filet mignon section. Broil.

loin tip   Lean, boneless steaks or roasts from the round. Broil or braise the steaks. Moist cook the roasts.

London broil   The only true London broil is the flank steak cut from the hindquarter. Lean cuts from the round and the chuck are also called London broil. Broil or braise.

lower round   Roast from the bottom or eye of round. Pot roast.

minute   Small, thin, steaks cut from the flank chuck or round. They are sometimes cubed. Pan fry.

minute sirloin   A very thinly cut steak from the short loin. Pan fry.

Newport rib   A semi-boned, easy-to-carve, first-cut rib roast. Roast.

New York   A short loin or shell steak without the tail end. Broil.

New York strip   Shell steak from the short loin. Broil.

New York style A fully trimmed, center-cut steak from the round. Broil or braise.

O.P. rib "Oven prepared" style rib roast used by restaurants. Roast.

outside bottom round Steaks and roast cut from the bottom and the eye of round. Pot roast the roasts. Pan fry the steaks.

outside round Steaks and roasts cut from the bottom round. Pot roast or pan fry.

plate A wasty primal cut generally used for grinding or moist cooking.

petite Small steak cut from a tender portion of the chuck. Pan broil or fry.

Pike's Peak A peak-shaped pot roast cut from the top and the heel of round. Moist cook.

pinbone sirloin A wasty steak cut from the loin end. Broil.

Porterhouse Best steak of the short loin and the entire beef carcass encompassing both the filet and strip portions and a bone. Broil.

regular Any steak or roast from the top of the round. Pan fry steak, roast or moist cook roast.

rib eye Small minute steak from the chuck section. Pan fry.

rolled cross rib A trimmed, rolled roast from the chuck. Moist cook.

rolled plate A rolled roast from the plate primal cut. Moist cook.

rolled rib Any fully trimmed rib roast that is rolled and tied. Roast.

rolled rump A lean roast from the hip end of the round. Pot roast or roast if of high quality.

round bone shank A cut from the foreshank or shin generally used for stews and stocks.

round muscle A less tender chuck roast. Moist cook.

round steak A cut from the round containing the top, bottom, and eye sections. Moist cook.

round tip A steak or roast from the side of round. Moist cook.

rump A lean roast from the tail section of the round. Moist cook or roast if of high quality.

sandwich A very thin cut of steak from the sirloin, round, or chuck sections. Pan broil.

Scotch tender A filet from the center section of the chuck for roasting.

seven-bone   A wasty cut from the chuck for steaks and roasts. Moist cook.

shell   A short loin steak that is the same as the club and T-bone without the tail ends. Broil.

short   A short loin (T-bone) steak. Broil.

short hip   A cut of beef from which all the sirloin steaks are cut. Broil the steaks.

short loin   The best section of the primal cut of beef from which the beef tenderloin roast, filet mignon, Porterhouse, T-bone, shell, and strip steaks are cut. Roast the roast. Broil the steaks.

short plate   Part of the primal cut of beef from which the skirt steak is cut. Most of the plate meat is used for grinding. Moist cook the steak.

short ribs   Bony part of the ends of rib and plate meat. Moist cook.

short sirloin   Small sirloin steaks from the middle of the sirloin section of the whole loin. Broil.

shoulder   Lean part of the chuck used for steaks and roasts. Moist cook.

shoulder clod   Boneless top chuck used for roasts and steaks. Moist cook.

silver side   A roast from the thick end of the top round. Roast.

silver tip   A subdivision of the side of round used for steaks and roasts. Broil or roast.

sirloin   A loin end or hip roast or steak. Roast the roast. Broil the steak.

sirloin butt   The whole side of round used for steaks and roasts. Pan fry or moist cook. Also loin end steak or roast. Broil or roast.

sirloin tip   Another name for the side of round that can be cut into roasts and steaks. Roast, pan broil, or moist cook.

skirt steak   The steak cut from the side of the plate section. Moist cook.

small T-bone   The first steaks cut from the small end of the short loin. Broil.

soup bone   Solid bones with good marrow from the foreshanks and hindshanks. Generally used for stocks and soups.

soup meat   Sinewy meat from the foreshank and hindshank.

Spencer A filet of the center section of the chuck used for roasts and steaks. Also a boneless rib steak or roast. Broil the steaks. Roast the roasts.

Spencer roll A fully trimmed and tied rib roast. Roast.

strip A boneless short loin (T-bone or shell) steak. Broil.

stuffed flank A flank steak into which a pocket has been inserted for stuffing. Moist cook.

Swiss Lean round steaks, often available cubed or scored with a knife and cut into individual portions. Braise.

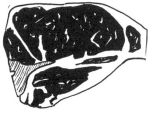

T-bone Steak of the short loin section—the same as a Porterhouse with less of the tenderloin section. Broil.

tender Trimmed, boneless steaks from the sirloin. Broil.

tenderette Cubed, individual round steaks. Pan broil.

tenderloin The excellent filet mignon roast from the short loin section. Roast.

thick rib Chuck roast cut closest to and shaped like a rib roast. Roast.

top butt The boneless, tender center of the sirloin section used for steaks and roasts. Broil or roast.

top loin A section of the loin without the filet. May be divided into roasts or steaks. Broil or roast.

top of chuck A chuck steak. Pan broil.

top of Iowa A fully trimmed sirloin steak. Broil.

top round The best section of the whole round for both steaks and roasts. Broil the steaks. Roast the roasts.

top sirloin The lean and tender side portion of the round used for steaks and roasts. Broil the steaks. Roast the roasts.

treasure A fully trimmed rib steak. Broil.

tripe The first and second stomachs of beef cattle. Moist cook.

turn-bone sirloin A sirloin steak encompassing several bones of varying sizes. Broil.

upper round   Steaks and roasts cut from the top section of the round. Fry or moist cook.

veiny   Another name for the side of round section from which steaks and roasts are cut. Broil or pan broil the steaks. Roast the roasts.

wedge bone   Small, end-cut steaks from the sirloin. They contain a large bone and a good deal of waste. Broil.

LAMB

American leg   Center-cut leg roast.

arm chops   A less tender shoulder-cut chop. Broil, pan broil, or braise.

blade chops   Less tender shoulder-cut chops. Broil, pan broil, or braise.

breast   Bony cut for roasting or braising. Sometimes sold boneless and served stuffed.

brisket pieces   Boneless cuts from the breast for moist cooking.

center leg   Roast cut from the center of the leg.

combination leg   Whole leg cut into halves—one for roasting, the other subdivided into steaks for pan frying, broiling, or braising.

crown roast   Fancy cut shaped from the rack section. Roast.

cushion shoulder   Boneless shoulder roast. Roast or pot roast.

English chops   Boneless loin chop filet encompassing part of the kidney. Pan broil, broil, braise.

Frenched rib chops   Rib section chop with meat scraped from the lower part of the rib bone. Sometimes served garnished with a paper frill. Broil, pan broil, braise.

kabobs   Cubed lamb usually cut from the leg or shoulder. Broil or barbecue.

lamburgers   Ground lamb from any primal section that has been formed into patties for pan frying.

leg   Primal cut which is sold whole or in halves for roasting, or sub-divided into chops and steaks for broiling, pan broiling, or braising.

leg chops   Center cut from leg. Broil or pan fry.

leg steaks   The same as leg chops. Broil or pan fry.

loin chops   Extremely desirable chops cut from the loin section. Broil, pan fry.

loin roast   Expensive primal cut of lamb for roasting.

neck   A sinewy cut from the tip of the shoulder used in recipes which call for moist cooking.

rack   The rib section of lamb from which chops and roasts are cut. Also sold whole or in sections for roasting.

rack roast   Roast of lamb from the rib section.

rib chops   High-quality chop from the rack section. Broil or pan fry.

rib roast   One half of the whole rack for roasting.

riblets   Small cuts from the breast of lamb, each containing a bone. Moist cook.

rolled breast   Boneless breast meat for roasting.

rolled double loin   Two de-boned loins rolled and tied together for roasting.

rolled leg   De-boned leg, rolled and tied for easy carving. Roast.

rolled shoulder   De-boned shoulder, rolled, shaped, and tied. Roast or pot roast.

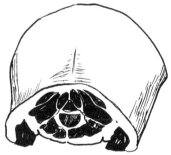

saddle   Both sides of the loin for roasting.

Saratoga chops   De-boned, rolled shoulder chops. Broil, pan fry, or barbecue.

shank   The extreme end of the leg for moist cooking.

shank half of leg   The easiest half of the leg to carve. Roast.

sirloin chops   Chop cut from the rump end of the loin. Broil or pan fry.

sirloin half of leg   The most difficult half of the leg to carve. Roast.

sirloin roast   Roast cut from the rump end of the loin.

square cut shoulder   Untrimmed shoulder roast. Roast or pot roast.

PORK AND HAM

arm roast of ham   Partially trimmed roast from the shoulder.

arm steak   Steak cut from the shoulder. Braise or pan fry.

back ribs   Shoulder end backbones of the loin section, also known as country-style spareribs. Bake, barbecue, or braise.

bacon   Extremely fatty smoked and cured cut generally from the belly but also from the jowl. Fry, bake, or broil.

blade chops   Chops from the shoulder butt section. Braise or pan fry.

blade loin roast   Roast from the shoulder end of the loin.

blade steaks   Steaks cut from the shoulder butt. Braise or pan fry.

boneless smoked shoulder butt   A de-boned, smoked version of the shoulder butt. Roast.

Boston butt   The shoulder butt section. Roast.

butt   The primal cut on top of the shoulder. Otherwise referred to as Boston butt.

Calia ham   Pork shoulder roast.

Canadian style bacon   Bacon cured from the tenderloin of pork. Fry, pan broil, or broil.

center-cut rib roast   Choice section of the rib end of the loin. Roast.

center loin chop   The most desirable pork chop cut from the center of the loin. Braise or pan fry.

center loin roast   Choice section of the full loin. Roast.

center rib chop   Chop from the center of the rib end of the loin. Braise.

chitterlings   Small intestine of pork. Soak for 6 hours, drain, and clean. Simmer in water and then deep fry.

country style spareribs   The back bones cut from the shoulder end of pork loin. Bake, barbecue, or braise.

cutlets   Small boneless cuts from the shank end of the fresh ham. Braise.

end-cut pork chops   Chops from either end of the loin. Less desirable than center cuts. Braise.

fatback   Fat covering the loin section used for flavoring and shortening.

foreleg   The pig's foot. Can be purchased fresh or pickled.

fresh ham   The haunch of the leg of pork. Roast.

fresh ham steak   Center slice of whole fresh ham. Pan fry, braise.

fresh picnic   Shoulder of pork. Roast.

fresh pork butt   Also known as the Boston butt or shoulder butt. Roast.

full loin roast   The entire loin end of the loin of pork.

full rib roast   The entire rib end of the loin of pork.

ham butt   Smoked shoulder of pork. Roast.

ham steak   Center-cut slices of the smoked, cured fresh ham. Braise, bake, or pan fry.

hocks   Slices for the extreme end of the leg of pork. They are braised, used in flavoring of vegetables, soups, and stews. They may be purchased either fresh or smoked.

jowl bacon   Bacon cut from the head of the hog. Fry, pan broil, broil, or use for flavoring other foods.

lard   Rendered pork fat, popular as a shortening.

loin chops   Excellent chops containing a section of the tenderloin cut from the rear section of the loin. Braise or pan fry.

picnic ham   A roast from the pork shoulder.

pig's feet   The tip end of the leg of pork. Can be purchased fresh or pickled. Braise or use for flavoring other foods.

pig's knuckles   Part of the pig's feet. Can be purchased fresh or pickled.

porklet   A cubed pork "cutlet" from the shoulder. Pan fry.

rib chops   Chops from the rib end of the loin. Braise or pan fry.

rolled Boston butt   Boneless, rolled roast from the shoulder butt.

rolled fresh ham   A boneless, rolled leg roast.

rolled loin roast   A boneless, rolled loin of pork.

salt pork   Cured cut from the belly of the hog, used primarily for flavoring and shortening.

slab bacon   Unsliced bacon, generally cut from the belly.

shoulder butt   A distinctively flavored but wasty roast.

shoulder of pork   The butt section of the hog for roasting.

shoulder steaks   Arm cuts ideal for barbecuing.

sirloin chops   Chops cut from the end of the pork loin. Braise or pan fry.

sirloin roast   Roast from the loin of pork.

smoked butt   A boneless, smoked roast from the shoulder of pork.

smoked ham   A term used to describe any section of pork that has been subjected to the smoking process.

smoked shoulder   A smoked pork shoulder for roasting.
spareribs   Bony cut from the breast of hog. Bake, braise, or barbecue.
streak o' lean   Another name for bacon.
tenderloin   The boneless loin of pork for roasting.
top loin chop   Center-cut loin chop. Braise or pan fry.

VEAL

arm roast   A cut from the shoulder (or chuck) of veal. Usually sold de-boned, rolled, and tied. Roast with added fat or braise.
arm steak   A slice from the shoulder of veal. Braise or pan fry.
birds   Thin stuffed slices of veal scallop. May either be roasted or braised.

blade roast   A roast from the shoulder of veal. Pan fry.

breast   A bony primal cut which may be purchased de-boned or with a pocket for stuffing. Roast or braise.
brisket pieces   Small pieces of breast of veal suitable for braising and stewing.
calves' feet   A variety meat used for flavoring stocks and gelatins.

center leg roast   Choice roast for braising or roasting.

choplets Cubed veal formed into the shape of a chop.

city chicken Cubes of veal served kabob style for roasting.

crown roast Elegant and expensive cut from the rib section of veal.

cutlet Small, boneless, solid piece of veal generally cut from the leg but also from the rib, loin, and shoulder. Braise, pan fry, or sauté.

foreshank The front leg of veal. Braise, stew, cook in liquid.

Frenched rib chop Rib chop with the meat scraped from the lower section of the rib bone. Braise.

heel of round The end of the leg section of veal. Braise, stew, cook in liquid.

kidney chops Loin chops encompassing a section of the veal kidney. Braise.

loin chops Select chops from the loin encompassing a section of the tenderloin. Braise or pan fry.

loin roast Expensive loin section of veal. Roast or braise.

mock chicken legs Ground veal on a stick, fashioned to look like chicken. Pan fry.

neck Sinewy cut suitable for stewing.

patties Ground veal for pan frying.

rib chops Rack section chop. Braise or pan fry.

rib roast Roast cut from the rack section. Roast or braise.

riblets Cuts from the breast containing portions of the rib bones. Suitable for moist cooking.

rolled double sirloin roast A special cut: two boneless sirloins rolled and tied for braising or roasting.

rolled leg Boneless leg, rolled and tied for roasting or braising.

rolled loin roast Boneless rolled loin, sometimes served stuffed. Roast or braise.

rolled shoulder Boneless shoulder, rolled and tied for braising or roasting.

scallops Small, ultra-thin solid slices of boneless veal generally from the leg but also from the rib, loin, and shoulder. May be sautéed, pan fried, or braised.

shank half of the leg   A bone-in roast from the extreme end of the vealer's leg. Roast or braise.

shoulder   A less tender primal cut also called chuck. It is cut into steaks, roasts, cutlets, scallops, and stew meat.

sirloin roast   A roast from the end of the loin section of veal. Roast, braise.

sirloin steak   A steak cut from the sirloin roast. Braise or pan fry.

standing rump   The rump end of the leg roast. Roast or braise.

4

A MEAT CALORIE COUNTER

Cut or Type of Meat	Description and/or Size of Serving	Number of Calories
Beef	USDA Choice—lean only	
beef, corned	medium fat, 4 oz.	424
beef, dried	4 oz.	231
brisket	4 oz.	253
chuck roast	4 oz.	220
flank steak	4 oz.	223
hamburger	lean, broiled, 4 oz.	250
Porterhouse steak	4 oz.	255
rib roast	4 oz.	275
round roast	4 oz.	205
sirloin steak	4 oz.	274
T-bone steak	4 oz.	254
Pork		
bacon	1 slice	48
Canadian bacon	1 slice	58
fresh ham	lean and fat, 4 oz.	426
ham: cured	lean only, 4 oz.	213
ham: boiled	1 slice, 4½″ by 4″ by ¼″	200
loin	lean, 4 oz.	290
pig's foot	½ pickled	125
rib chop	3″ by 5″ by 1″	290
roast	lean only, 4 oz.	279

CUT OR TYPE OF MEAT	DESCRIPTION AND/OR SIZE OF SERVING	NUMBER OF CALORIES
spareribs	6 average	246
tenderloin	2 oz.	200
Lamb	USDA Choice—lean only	
leg roast	4 oz.	212
loin chop	4.8 oz. with bone	140
shoulder roast	4 oz.	234
stew	1 cup, with vegetables	250
Veal		
breaded cutlet	½ cutlet	280
chuck	moist cooked, 4 oz.	266
loin chop	⅝″ thick	225
plate	moist cooked, 4 oz.	344
roast	2 slices, 3″ by 2″ by ⅛″	186
Variety Meats		
brains	4 oz.	142
heart	beef, 4 oz.	214
heart	calf, 4 oz.	237
kidney	beef, 4 oz.	287
liver	beef, 4 oz.	260
liver	calf, 4 oz.	296
sweetbreads	4 oz.	191
tongue	beef, 4 oz.	277
tongue	calf, 4 oz.	181
tripe	pickled, 4 oz.	70
Sausages		
blood sausage	4 oz.	447
bockwurst	4 oz.	301
bologna	1 slice	66
braunschweiger	4 oz.	364
cervelat	4 oz.	514
frankfurter	1 average	151
head cheese	4 oz.	305

CUT OR TYPE OF MEAT	DESCRIPTION AND/OR SIZE OF SERVING	NUMBER OF CALORIES
knackwurst	4 oz.	317
liverwurst	1 slice	80
pork links	3″ by ½″, each	94
salami	4 oz.	510
scrapple	4 oz.	244

Bibliography

BATES, MARSTON, *Gluttons and Libertines,* New York: Random House, 1967.

CROSS, JENNIFER, *The Supermarket Trap,* Bloomington: University of Indiana Press, 1970.

LEVIE, ALBERT, *The Meat Handbook,* Wesport, Conn.: The Avi Publishing Co., Inc.

MARGOLIUS, SIDNEY, *The Great American Food Hoax,* New York: Dell Publishing Co., 1972.

ROBERSON, JOHN AND MARIE, *The Meat Cookbook,* New York: Collier Books, 1966.

SIMOONS, FREDERICK J., *Eat Not This Flesh,* Madison: The University of Wisconsin Press, 1967.

TRAGER, JAMES, *The Enriched, Fortified, Concentrated, Country-Fresh, Lip-Smacking, Finger-Licking, International, Unexpurgated Foodbook,* New York: Grossman Publishers, 1970.

PAMPHLETS:

AMERICAN MEAT INSTITUTE, *Ideas With Meat,* Chicago.

———, *Sidelights on Sausage,* Chicago.

———, *The Hot Dog,* Chicago.

———, *The Story of Pork,* Chicago.

BURRIS, MARY ELLEN, AND BIESDORF, HEINZ B., *Be a Better Shopper,* Ithaca: Extension Publication of the New York State College of Human Ecology at Cornell University, 1969.

MOOLMAN, VALERIE (ed.), *How to Buy Food,* New York: Cornerstone Library, 1970.

NATIONAL LIVE STOCK AND MEAT BOARD, *Cooking Meat in Quantity,* Chicago.

————, *Facts About Sausages,* Chicago.

————, *Lessons on Meat,* Chicago.

SUPERMARKET INSTITUTE AND AMERICAN MEAT INSTITUTE, *Freezing and Handling of Meats in the Home,* Chicago, 1957.

U.S. DEPARTMENT OF AGRICULTURE, *How to Buy Beef Roasts,* Washington, D.C.

ARTICLE:

EDITORS OF *National Provisioner,* "Nutritional Labelling Is on the Way," April 29, 1972.

Index